T0267396

Principles of
Fire Risk Assessment in
Buildings

Principles of Fire Risk Assessment in Buildings

David Yung
Yung & Associates Inc., Toronto, Canada

A John Wiley and Sons, Ltd, Publication

Registered office
John Wiley & Sons Ltd, The Atrium, Southern Gate, Chichester, West Sussex, PO19 8SQ, United Kingdom

For details of our global editorial offices, for customer services and for information about how to apply for permission to reuse the copyright material in this book please see our website at www.wiley.com.

Disclaimer

Library of Congress Cataloging-in-Publication Data

Yung, David Tin Lam.
 Principles of fire risk assessment in buildings / David Tin Lam Yung.
  p. cm.
 Includes bibliographical references and index.
 ISBN 978-0-470-85402-0 (cloth) – ISBN 978-0-470-85409-9 (pbk. : alk. paper)
1. Fire risk assessment. I. Title.
 TH9446.3.Y86 2009
 363.37'6–dc22
                                                          2008043731

A catalogue record for this book is available from the British Library.

ISBN: 978-0-470-85402-0 (Hbk)
     978-0-470-85409-9 (Pbk)

Typeset in 10.5/13 Sabon by Laserwords Private Limited, Chennai, India

Contents

4   Qualitative Fire Risk Assessment   33
     4.1   Overview   33
     4.2   Risk Matrix   34
     4.3   Checklist Method   36
     4.4   Event-Tree Method   42
     4.5   Summary   46
     4.6   Review Questions   47
           References   47

5   Quantitative Fire Risk Assessment   49
     5.1   Overview   49
     5.2   Risk Indexing   50
     5.3   Checklist Method   50
     5.4   Event-Tree Method   55
     5.5   Summary   60
     5.6   Review Questions   61
           References   62

PART II   Fundamental Approach to Fire Risk Assessment   63

6   Fundamental Approach to Fire Risk Assessment   65

7   Fire Growth Scenarios   71
     7.1   Overview   71
     7.2   Compartment Fire Characteristics   72
     7.3   Fire Model Input and Output Parameters   76
     7.4   Design Fires   80
     7.5   Automatic Fire Suppression to Control Fire Growth   89
     7.6   Summary   91
     7.7   Review Questions   92
           References   93

8   Fire Spread Probabilities   95
     8.1   Overview   95
     8.2   Fire Resistant Construction   96
     8.3   Probability of Failure   100
     8.4   Fire Spread Probabilities   106
     8.5   Summary   110
     8.6   Review Questions   111
           References   112

About the Author

The author is currently the President of his own consulting company, Yung & Associates Inc. He has worked in fire research and fire risk assessment for over 20 years. From 2002 to 2006, he was Research Leader of Fire Science at the Australian national research organization CSIRO in Sydney, Australia. Before that, he was a Senior Research Officer and Group Leader of Fire Risk Assessment for 17 years at the National Research Council Canada (NRCC) in Ottawa. He led a team that developed one of the world's comprehensive fire risk-cost assessment models, called FiRECAM. Before joining the fire research group at NRCC in 1985, he spent seven years at the Argonne National Laboratory in the USA, and three years at the Chalk River Nuclear Laboratory and NRCC, conducting nuclear and solar energy research.

The author's work is mentioned in the book *History of Fire Protection Engineering*, published in 2003 by the National Fire Protection Association (NFPA) and the Society of Fire Protection Engineers (SFPE) in the USA. He is a member of the NFPA Technical Committee on Fire Risk Assessment Methods. In September 2003, he was given the 'Hats Off' award by SFPE for his service as the Editor-in-Chief of the *Journal of Fire Protection Engineering*. In May 2001, he was elected a Fellow of the SFPE for his achievement in fire protection engineering. He has over 100 publications in fire, nuclear and solar energy, and desalination and water purification. He serves on the editorial boards of three major international fire journals and is the past Editor-in-Chief of the *Journal of Fire Protection Engineering*.

He holds a B.Eng. from McGill University, an M.A.Sc. from the University of Toronto, and a Ph.D. from MIT, all in mechanical engineering. In addition, he is a licensed professional engineer in the Province of Ontario, Canada and a member of the American Society of Mechanical Engineers.

Preface

The concept for this book originated in late 2001 when John Wiley & Sons Inc. approached me with a proposal to write a book on fire risk assessment. At that time, I was a Senior Research Officer at the National Research Council Canada (NRCC) and had conducted research on fire risk assessment for 15 years. I was happy to accept the challenge to write a book based on my observations and research findings on fire risk assessment over the years.

Shortly after agreeing to write the book, I accepted an invitation from CSIRO in Australia to lead their fire science group. Working in a new organization and living in a new country, however, did not permit me much time to work on a book – hence, it wasn't until I returned to Canada in 2006 that writing began in earnest.

My involvement in fire risk research was preceded by a number of years in energy research. I had spent time at the Argonne National Laboratory in Chicago working on thermal hydraulic issues related to nuclear reactor safety, and heat transfer issues related to Ocean Thermal Energy Conversion in Hawaii. I also spent time at the Chalk River Nuclear Laboratory working on heat transfer issues related to the Canadian CANDU reactor; and later at NRCC conducting research in solar energy. The cessation of government funding for energy research in 1985 resulted in a career change to fire research, in particular fire risk assessment research.

My interest in fire risk assessment research was influenced by two individuals. The first was Professor Vaughn Beck of the Victoria University of Technology in Australia who spent time on sabbatical at NRCC in early 1987. His great enthusiasm for the development of computer-based fire risk assessment models was contagious. The second was Ken Richardson who was, at that time, the Associate Head of National Fire Laboratory at

NRCC. He saw the need for fire risk assessment models and encouraged me to lead research in that area. As a result, I formed a team at NRCC to conduct research and to develop fire risk assessment models. The team collaborated successfully with Professor Beck's Australian team throughout the 1990s. The NRCC team had been strengthened by the addition of two prominent researchers: Dr George Hadjisophocleous and Dr Guylene Proulx, both of whom made significant contributions to the development of fire risk assessment models at NRCC. Many of the concepts described in this book were developed by the NRCC and the Victoria University of Technology researchers throughout that period and I am grateful for all of their contributions.

This book has been prepared as a reference source for fire safety professionals working in the fire risk assessment field. It is also intended as a textbook for university students in fire protection engineering. It is my hope that it will serve both fields well.

David Yung
Toronto, Canada

Acknowledgments

The author would like to express his sincere thanks to the SFPE Educational & Scientific Foundation and the PLC Foundation in the USA for providing generous funding support to this work.

The author would also like to thank Mr Doug Brandes and Dr Dave Evans of the SFPE Educational & Scientific Foundation and Mr Ken Dungan of the PLC Foundation for their encouragement and support of this work.

The author is also indebted to his friend, Dr Yunlong Liu of Sydney Australia, for providing many of the computer model output illustrations.

List of Symbols

Alphabets

A_f	Floor area of the compartment in Equation 8.2 (m^2)
A_t	Total area of internal surfaces in Equation 8.2 (m^2)
A_v	Total area of openings in the walls in Equation 8.2 (m^2)
c_p	Specific heat ($J \cdot kg^{-1} \cdot K^{-1}$)
C	Consequence
C_l	Negative coefficient on the leeward side in Equation 9.7
C_w	Positive coefficient on the windward side in Equation 9.7
CC	Capital cost (\$)
D	Optical Density (m^{-1}) or crowd density (persons$\cdot$m^{-2})
e_f	Fuel load density in Equation 8.1 ($MJ \cdot m^{-2}$)
EFC	Expected fire cost (\$)
ERL	Expected risk to life (deaths)
F	Occupant flow rate per unit width of the evacuation route (persons$\cdot$s$^{-1} \cdot$m^{-1})
FID	Fractional incapacitating dose
FRR	Fire resistance rating (min or hr)
g	Gravitational acceleration ($m \cdot s^{-2}$)
H_v	Height of the openings in Equation 8.2 (m)
ΔH_c	Maximum heat of combustion per unit fuel burned ($kJ \cdot g^{-1}$)

ΔH_{eff}	Effective heat of combustion per unit fuel burned $(kJ \cdot g^{-1})$
ΔH_{rad}	Flame radiant loss per unit fuel burned $(kJ \cdot g^{-1})$
k	Thermal conductivity $(W \cdot m^{-1} \cdot K^{-1})$ or characteristic occupant travel speed $(m \cdot s^{-1})$
k_c	Compartment boundary parameter in Equation 8.1 $(min \cdot m^{2.25} \cdot MJ^{-1})$
K_m	Specific extinction coefficient $(m^2 \cdot g^{-1})$
L	Fire load in Equation 12.10 (min)
L_w	Latent heat of vaporization of water $(MW \cdot L^{-1} \cdot s^{-1})$
$\dot{m}$	Mass flow rate $(kg \cdot s^{-1})$ or pyrolysis rate $(g \cdot s^{-1})$
MC	Maintenance cost (\$)
OCC_{TR}	Number of trapped occupants
OF_i	Occupant fatalities for fire scenario i (deaths)
OF_{NR}	Additional occupant fatalities as a result of no rescue by firefighters (deaths)
OF_{SS}	Occupant fatalities as a result of smoke hazard (deaths)
p	Pressure $(N \cdot m^{-2})$ or probability distribution
p_i	Indoor pressure $(N \cdot m^{-2})$
p_l	Wind pressure on the leeward side of a building $(N \cdot m^{-2})$
p_o	Outdoor pressure $(N \cdot m^{-2})$
p_s	Standard pressure at sea level $(N \cdot m^{-2})$
p_w	Wind pressure on the windward side of a building $(N \cdot m^{-2})$
P_{DV}	Probability of success of closing doors and shutting off ventilation systems
P_{FD}	Probability of ignition developing into a fully-developed fire
P_{FO}	Probability of flashover fires
P_{FS}	Probability of fire spread
P_i	Probability of occurrence of fire scenario i over the design life of the building
P_{IG}	Probability of ignition
P_{NF}	Probability of non-flashover fires
P_{SC}	Probability of success of extracting smoke on the fire floor and pressurizing the floors above and below
P_{SM}	Probability of smouldering fires
P_{SP}	Probability of success of pressurizing stairwells and elevator shafts

P_{SS}	Probability of smoke hazard
P_{EXT}	Firefighter's extinguishment effectiveness
P_{RES}	Firefighter's rescue effectiveness
P_{SFO}	Probability of suppressing flashover fires
P_{SNF}	Probability of suppressing non-flashover fires
P_{SSM}	Probability of suppressing smouldering fires
PL_i	Property loss for fire scenario i ($)
PIT	Probability of incapacitation from temperature
PRO	Property value ($)
$P(call)_i$	Probability of calling the fire department at fire State i
$P(det)_i$	Probability of fire detection at fire state i
$P(firstcall)_i$	Probability of notifying the fire department for the very first time at fire State i
$P(occu)_i$	Probability of occupants calling the fire department at fire State i
$P(FRR)$	Cumulative probability of probable fires up to FRR in Equation 8.3
$P'(FRR)$	Probability of failure of FRR in Equation 8.4
$P_R[C_i]$	Reliability of component C_i
Q_c	Convective heat loss through the openings (kJ)
$\dot{Q}$	Heat release rate (kW)
R	Fire resistance in Equation 12.10 (min or hr)
R_a	Specific gas constant for air $(N{\cdot}m{\cdot}kg^{-1}{\cdot}K^{-1})$
R_{FO}	Ratio of firefighter's intervention time to the flashover time
RFL_W	Required water flow rate to absorb the heat release rate of the fire $(L{\cdot}s^{-1})$
S	Occupant travel speed $(m{\cdot}s^{-1})$, or design safety margin in Equation 10.12 (min)
t	Time (s)
t_e	Equivalent time of the ISO 834 standard fire in Equation 8.1 (min)
T_a	Ambient temperature (K)
T_g	Smoke temperature (K)
T_i	Indoor temperature (K)
T_o	Outdoor temperature (K)
u	Velocity $(m{\cdot}s^{-1})$
u_i	Hasofer-Lind transformed parameters in a multi-dimensional standard space
V	Volume (m^3)

w	Ventilation factor in Equation 8.1 ($m^{-0.25}$)
x_i	Controlling parameters in a multi-dimensional space
y_{CO}	CO yield per unit fuel burned (g/g)
y_{CO2}	CO_2 yield per unit fuel burned (g/g)
y_s	Soot yield per unit fuel burned (g/g)
z	Vertical coordinate (m)
z_n	Neutral plane height (m)

Greek Symbols

α	Fire growth rate coefficient in Equation 7.1 ($kW{\cdot}s^{-2}$)
λ_i	Failure rate of component i (hr^{-1})
μ	Mean
η_w	Efficiency of water application by firefighters to absorb the heat from a fire
ρ	Density ($kg{\cdot}mg^{-3}$)
ρ_a	Air density ($kg{\cdot}m^{-3}$)
ρ_g	Smoke density ($kg{\cdot}m^{-3}$)
ρ_i	Indoor air density ($kg{\cdot}m^{-3}$)
ρ_o	Outdoor air density ($kg{\cdot}m^{-3}$)
ρ_s	Smoke mass density in Equation 10.3 ($g{\cdot}m^{-3}$)
σ	Standard deviation

1

Introduction

The practice of fire safety designs is changing in many countries. The change is from traditional practice that simply follows the prescriptive code requirements to those that are based on fire safety analysis to obtain the required level of fire safety for the occupants. The change is a result of many countries moving towards the more flexible performance-based codes. Performance-based codes allow flexibility in fire safety designs as long as the designs can provide the required level of fire safety to the occupants.

Fire risk assessment is an assessment of the fire risks, or the levels of fire safety, that are provided to the occupants and property in a performance-based fire safety design. Fire safety designs involve the use of fire protection measures to control fire growth and smoke spread and to expedite occupant evacuation and fire department response. None of these fire protection measures, however, is 100 % effective. For example, sprinklers do not have 100 % reliability in controlling fires, nor do fire alarms have 100 % reliability in getting occupants to leave immediately. As a result, certain levels of fire risks to the occupants and property are implied in each fire safety design. The assessment of these levels of fire risks is the subject of fire risk assessment.

Guidelines on fire risk assessment have been produced by fire protection organizations such as the NFPA (National Fire Protection Association) and SFPE (Society of Fire Protection Engineers) in the USA (NFPA 551, 2007; SFPE, 2006). Other international organizations such as ISO are also planning to introduce reference documents on fire risk assessment. These guidelines are for the benefits of fire protection engineers and regulators to allow them to have a common vision on what is required in the submission and approval process in fire risk assessment.

Principles of Fire Risk Assessment in Buildings   D. Yung
© 2008 John Wiley & Sons, Ltd

They describe this process from beginning to end, including the setting of risk thresholds and the selection of fire scenarios. These guidelines, however, do not describe the actual fire risk analysis. This book, on the other hand, describes the basic principles of fire risk analysis, or fire risk assessment, in buildings. This book, therefore, is suitable for use as a reference to these other guidelines.

Research and technical papers are produced regularly on the advancement of fire risk assessment. These papers usually focus on a certain aspect of the fire risk assessment. They seldom describe the fundamentals that underpin fire risk assessment. This book is suitable for use as a reference to these papers.

This book is also suitable for use as a textbook on fire risk assessment. The book describes the complex fire risk assessment principles in a way that is easy to follow.

This book is divided into two parts. The first part is devoted to the traditional fire risk assessment methods. The first part consists of four chapters, from Chapter 2 to Chapter 5. The second part is devoted to fire risk assessment methods based on a fundamental approach. The second part consists of eight chapters, from Chapter 6 to Chapter 13.

Chapter 2 is an introduction to fire risk assessment. Fire protection measures are shown as fire barriers. They are grouped into five major barriers. The risks to occupants and property depend on how successful these barriers are in controlling fire initiation, fire growth, smoke spread, and in expediting occupant evacuation and fire department response.

Chapter 3 is a discussion of how fire risk assessment can be conducted by using past experience or incident data. This approach is only valid if the present situation and those in the past are exactly the same. Often, they are not.

Chapter 4 is a discussion of how qualitative fire risk assessment is conducted. Qualitative fire risk assessment involves the use of risk matrix, checklist method or event tree, and the use of qualitative subjective opinion on the occurrence and consequence of fire hazards.

Chapter 5 is a discussion of how quantitative fire risk assessment is conducted. Quantitative fire risk assessment involves also the use of risk matrix, checklist method or event tree, and the use of quantitative subjective opinion on the occurrence and consequence of fire hazards.

Chapter 6 is an introduction to fire risk assessment based on a fundamental approach. Fire scenarios are constructed based on the success and failure of fire protection measures. For each fire scenario, the outcome of occupant deaths and property loss is determined based on modelling of fire growth, smoke spread, occupant evacuation, fire department response and eventually fire spread through breaching boundary elements. The assessment of risks to life and property is based on occupant deaths and property losses from all fire scenarios.

Chapter 7 is a discussion of fire growth scenarios. Fire growth scenarios are constructed based on the success and failure of fire control measures. The fundamental characteristics of fire growth in a compartment are described. The development of a fire in the compartment of fire origin can be modelled using fire growth models.

Chapter 8 is a discussion of fire spread probabilities. The probability of failure of a boundary element is described. The probability of fire spread through multiple boundary elements is also described. Fire spread through multiple fire resistant boundary elements is a relative slow process in comparison to smoke spread, occupant evacuation and fire department response.

Chapter 9 is a discussion of smoke spread scenarios. Smoke spread scenarios are constructed based on the success and failure of smoke control measures. The fundamental characteristics of smoke spread are described. Smoke spread in a building can be modelled using smoke spread models.

Chapter 10 is a discussion of occupant evacuation scenarios. Occupant evacuation scenarios are constructed based on the success and failure of occupant evacuation measures. The fundamental characteristics of occupant evacuation are described. Occupant evacuation can be modelled using occupant evacuation models. Early evacuation is critical. Occupants are trapped in the building if they can not evacuate in time before the arrival of the critical smoke conditions in the evacuation routes that prevent evacuation.

Chapter 11 is a discussion of fire department response. The fundamental characteristics of fire department response are described. The effectiveness of fire department rescue and suppression efforts depends on fast response time and adequate resources. For occupants who are trapped and can not be rescued by firefighters, expected deaths are assessed based on the length of their exposure to untenable smoke and fire conditions.

Chapter 12 is a discussion of uncertainty in fire risk assessment. Probability concepts are introduced. The discussion is mainly on uncertainty, or reliability, in fire safety designs. Methods that can be used to assess uncertainty are described.

Chapter 13 is a discussion of fire risk management. Fire risk management includes the consideration of cost-effective fire safety design options that can provide equivalent level of fire safety but have the lowest fire costs. Fire risk management also includes the consideration of regular inspection and maintenance of fire protection systems to ensure that these systems can maintain their reliabilities. Some previous case studies from the computer fire risk-cost assessment model *FiRECAM* are discussed.

References

NFPA 551 (2007) *Guide for the Evaluation of Fire Risk Assessments*, National Fire Protection Association, Quincy, MA.
SFPE (2006) *Engineering Guide to Fire Risk Assessment*, Society of Fire Protection Engineers, Bethesda, MD.

Part I

Simple Approach to Fire Risk Assessment

2

What is Fire Risk Assessment?

2.1 Overview

Basic concepts of fire risk assessment will be introduced in this chapter. The term fire risk assessment refers to assessing risks to both people and property as a consequence of unwanted fires. In a simple risk assessment the probability of a certain unwanted fire scenario is considered and the consequence of that scenario are explored. In a comprehensive risk assessment all probable unwanted fire scenarios and their consequences are considered.

A fire scenario involves the projection of a set of fire events, all of which are linked together by whether the fire protection measures succeed or fail. The probability of a fire scenario is dependent on the individual probabilities of success or failure of fire protection measures. The risk to the occupants depends not only on the probability of the fire scenario that can lead to harm to the occupants, but also the level of harm to the occupants as a result of the consequence of that scenario. The consequence of a fire scenario can be assessed by using time-dependent modelling of fire and smoke spread, occupant evacuation and fire department response.

2.2 What is Fire Risk Assessment?

Fire risk assessment is the assessment of the risks to the people and property as a result of unwanted fires. It employs the same basic principles of risk assessment that are used in many other fields. A simple risk assessment considers the probability of the occurrence of a certain unwanted fire scenario and the consequence of that scenario. A comprehensive risk assessment considers all probable unwanted fire scenarios and their consequences. The definition of fire scenario will be discussed in the next section. It involves the linking of anticipated fire events by the success or failure of certain fire protection measures.

Consider, as an example, the assessment of the expected risk to life to the occupants in a building as a result of one single fire scenario. The expected risk to life can be expressed by the following equation:

$$\text{Expected risk to life} = P \cdot C, \qquad (2.1)$$

where P is the probability of a certain fire scenario and C is the expected number of deaths as a consequence of that fire scenario. If the probability of a certain fire scenario occurring in a building is once every 20 years, then $P = 0.05$ fires per year. If the consequence of that fire scenario is two deaths, then $C = 2$ deaths per fire. From Equation 2.1, the expected risk to life as a result of that fire scenario is equal to 0.1 deaths per year, or 1 death every 10 years.

Because fires can occur in a building in more ways than one, the risk to the occupants is usually assessed based on all probable fire scenarios. A comprehensive fire risk assessment can be expressed by the following equation:

$$\text{Expected risk to life} = \sum_i (P_i \cdot C_i), \qquad (2.2)$$

where Σ represents the summation of all probable fire scenarios, P_i is the probability of one fire scenario, i, and C_i is the expected number of deaths as a consequence of that fire scenario, i.

It should be noted that fire risk assessments involve more than the assessment of the risk to life. It involves also the assessment of the loss of property, loss of business and so on, as a result of fires. Equations similar to Equation 2.1 and 2.2 can also be expressed for the other losses.

2.2.1 Fire Scenarios

A *fire scenario* is a sequential set of fire events that are linked together by the success or failure of certain fire protection measures. A *fire event* is an occurrence that is related to fire initiation, or fire growth, or smoke spread, or occupant evacuation, or fire department response. For example, a fire event can be: a fire develops into a post-flashover fire, or the occupants can not evacuate quickly enough and are trapped in the building, or the fire department responds in time and rescues the trapped occupants. A *fire protection measure* is a measure that can be a fire protection system, such as sprinklers and alarms; or a fire protection action, such as occupant evacuation training and drills.

A simple example of a fire scenario is the following set of events that are linked together by the failure of fire protection measures: a fire develops into a post-flashover fire, the alarm system does not activate and the occupants receive no warning signals and are trapped in the building. Another simple example is the following set of events that are linked together by the success of fire protection measures: a fire does not develop into a post-flashover fire, the alarm system activates, and the occupants receive the warning signals and evacuate the building. In real-world fires, fire scenarios are much more complex and the possible number of fire scenarios can be many. The number of fire scenarios depends on the number of permutations that can be constructed based on all the fire protection measures that are in place and all the fire events that are anticipated. The proper construction of fire scenarios and the proper analysis of the consequence of the fire scenarios, however, are the key to a credible fire risk assessment.

The general principle of how fire scenarios can be constructed will be discussed in the next section. But before we discuss how they can be constructed, take the simple case where there is no fire protection measure at all. Take, for example, the case where a fire occurs at the only exit door in a room with a number of people inside the room. With no fire protection in the room to control the fire, the fire develops into a post-flashover fire and subsequently kills all the people in the room. The risk of this simple fire scenario is the probability of a fire occurring at the only exit door of a room, multiplied by the number of people killed by this fire. Obviously, fire risk assessment is not as simple as this.

There are normally fire protection measures in place to protect the occupants and property. For example, there are normally fire protection measures in place to control the development of a fire and also to prevent the fire from spreading to other parts of a building. There are also normally fire protection measures in place to provide early warnings to the people and to help the people to get to a safe place before the fire spreads. There are also expectations that the fire department is notified and that they will come to extinguish the fire and rescue the people. Hence, fire risk assessment involves the use of fire scenarios that are

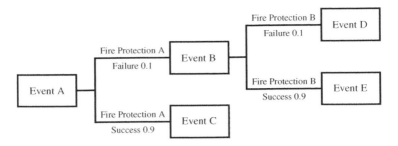

Figure 2.1 A simple event tree where an initiating event can lead to different events depending on the success and failure of fire protection measures at the branch points.

based on the success and failure of these fire protection measures in order to assess the expected risks to the occupants and the property.

A set of fire scenarios can be constructed based on the well-known *event-tree* concept, where events are linked together like the branches of a tree (Custer and Meacham, 1997). Figure 2.1 shows a simple event tree where an initiating event can lead to different events depending on the success or failure of the fire protection measures at the branch points. For example, Event A terminates in Event C if the fire protection measure for that event succeeds, whereas Event A continues with Event B to others if the fire protection measure fails. A particular set of events that are linked together forms one fire scenario. For example, the set of Event A and Event C forms one scenario. A set of all possible combinations of the linked events forms a complete set of all possible fire scenarios. For example, the combinations of A–C, A–B–D and A–B–E form a complete set of three fire scenarios.

Figure 2.1 also shows the probability of success or failure of these two fire protection measures at the two branch points. The probabilities of failure at the two branch points are assumed, for this example, to be the same, at 10 % or 0.1. Based on this, Scenario A–C has a probability of 0.9. Scenario A–B–E has a probability of 0.09, obtained by multiplying the probability of A–B (0.1) and that of B–E (0.9). Similarly, Scenario A–B–D has a probability of 0.01. The combined probability of all three fire scenarios is one. The important thing to note here is that the probabilities of success or failure of fire protection measures affect the probabilities of all fire scenarios. The lower the probabilities of failure of fire protection measures, the lower the probabilities of all those fire scenarios that will lead to an undesirable outcome. For example, if Event D is not the desired end point, then lower probabilities of failure of fire protection measures will lead to a lower probability of the undesirable fire Scenario A–B–D. If the probabilities of failure of the two fire protection measures are reduced to 0.01, the probability of the undesirable Scenario A–B–D is reduced to 0.0001.

2.2.2 Fire Protection Measures as Fire Barriers

For fire risk assessments in buildings, the event tree can be constructed based on the following five major fire events. They are considered major events because each is related to a major phase of fire development and hazard: fire ignition, fire growth, smoke spread, failure of occupants to evacuate, and failure of fire department to respond (Yung and Benichou, 2003).

1. *Fire ignition* is the initiating event, such as cigarette ignition of a couch in a living room or a mattress in a bedroom. Fire protection measures include fire prevention education, or the use of fire-retarded material in furniture, which would help to reduce the probability of occurrence of this event and the consequential risks.
2. *Fire growth* is the second event, which includes various types of fire growths, from fires developing into smouldering fires to fires developing into post-flashover fires. Fire protection measures include sprinklers, compartmentation and door self-closers, which would help to contain these fires and reduce their consequential risks. The reduction in risk depends on the reliability and effectiveness of these fire control systems.
3. *Smoke spread* to critical egress routes and other locations in a building is the third event. Fire protection measures include door self-closers, smoke control, and stairwell pressurization, which would help to contain the smoke and reduce its consequential risks. The reduction in risk depends on the reliability and effectiveness of these smoke control systems.
4. *Failure of occupants to evacuate* as a result of the spread of fire and smoke to egress routes is the fourth event. Fire protection measures include smoke alarms, voice communication, protected egress routes, refuge areas, and evacuation training and drills, which would help to provide early warnings to occupants, safe egress routes, quick occupant response and evacuation to either exit the building or to seek temporary protection in refuge areas. The reduction in risk depends on the reliability and effectiveness of these early warning and evacuation systems and the implementation of regular occupant training and evacuation drills.

5. *Failure of fire department to respond* in time to rescue any trapped occupants and control the fire is the fifth event. Protection measures include early fire department notification and adequate fire department resources. The reduction in risk depends on the reliability of early notification and adequacy of fire department resources.

Except for the occupants in the room of fire origin, all of the above five major hazardous events must occur before a fire can cause harm to the occupants. Each of the five hazardous events, however, can only happen if the fire protection measure for that event fails to prevent that event from happening. The fire protection measure for each event, therefore, can be viewed as a major *barrier* to that event. Potentially, there can be five major barriers between a fire and the people, as depicted in Figure 2.2. The barrier to prevent failure of occupant to evacuate is to facilitate occupant evacuation. The barrier to prevent failure of fire department to respond is to facilitate fire department response.

It should be noted that each major barrier represents a group of individual barriers, each of which can provide the same fire protection. For example, a major barrier to fire growth can consist of sprinklers, fire resistant compartmentation and door self-closers. Obviously, not all of these fire barriers are necessarily put in place in any building. How many are put in place depends on how many are required by the building regulations and how well the fire protection design is. But the more they are put in place, the better is the protection. Also, the more effective the barrier is, the better is the protection.

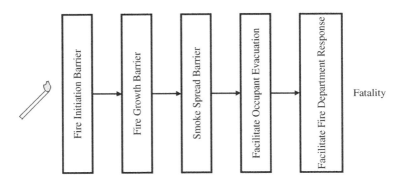

Figure 2.2 Five major fire barriers between a fire source and fatality (from Yung and Benichou, 2003, reproduced by permission of the National Research Council Canada).

The risk to the occupants depends, as discussed in the previous section, on the probability of failure of all fire protection measures, or barriers. For example, if there are no barriers at all, there is no protection. The probability of the fire scenario that can lead to harm to the occupants is 100 %, or 1. If there are five barriers and each barrier has a probability of failure of 0.5, the probability of failure of all five barriers is $0.5 \times 0.5 \times 0.5 \times 0.5 \times 0.5$ or 0.03125. The probability of the fire scenario that can lead to harm to the occupants is 0.03125. The risk to the occupants depends not only on the probability of the fire scenario, but also the actual harm to the occupants as a result of the consequence of that scenario (see Equation 2.2). The consequence of that fire scenario depends on how fast the fire and smoke spread in the building and how quickly the occupants evacuate the building, which will be discussed briefly in the next section.

The assessment of the probabilities, consequences and risks of fire scenarios will be the main focus in later chapters of this book. In here, we will look at the effect of the number of barriers and the reliability of the barriers on the probability of the fire scenario that can lead to harm to the occupants. We have discussed previously the case of five barriers with each having a probability of failure of 0.5. The probability of failure of all five barriers is 0.03125 and hence the probability of the fire scenario that can lead to harm to the occupants is 0.03125. Let us look at another case when there are only two barriers but with each having a lower probability of failure of 0.1. The probability of failure of both barriers is 0.01. The probability of the fire scenario that can lead to harm to the occupants is therefore 0.01, which is less than that of the previous case when there were five barriers but with each having a higher probability of failure of 0.5. This is the reason why fire risk assessment concerns not only the number of fire protection measures that are put in place, but also how reliable and effective these fire protection measures are. This also explains why the use of redundancy helps to increase the reliability of fire protection measures. For example, two fire protection measures with an individual probability of failure of 0.1 would provide a combined probability of failure of 0.01, which is less than that of one fire protection measure alone.

2.2.3 Time Factor in Consequence Modelling

As was discussed in the previous section, fire barriers help to reduce fire risks in two different ways: (1) control the development of a fire in the location of fire origin and its spread to other locations; and

(2) expedite the evacuation of the occupants and the response of the fire department. Barriers 1, 2 and 3 are those that try to control the development and spread of a fire; whereas Barriers 4 and 5 are those that try to expedite the evacuation and rescue efforts. Figure 2.2 shows that the evacuation of the occupants is as important as the control of the fire in fire risk assessment. If occupants can get to a safe place before hazardous conditions get to the egress routes, there is no risk to the occupants.

Fire barriers help to reduce the probabilities of those fire scenarios that can lead to harm (see Figure 2.1). The more effective the fire barriers are, the lower the probabilities of those fire scenarios that can lead to harm, and consequently the risk to the occupants. The consequence of these fire scenarios can be assessed by using time-dependent modelling of fire and smoke spread, occupant evacuation and fire department response, under conditions that are specific to each fire scenario. The basic principles of time-dependent modelling of fire scenarios will be described in later chapters. But before we do that, we will use a simple example here of a house fire to look at fire barriers and why time-dependent modelling is important in the assessment of the consequences of fire scenarios.

Based on Canadian fire statistics, the most frequent fatal fire scenario in house fires is the one that involves the ignition of a couch in the living room (Yung and Lougheed, 2001). That means Barrier 1 is not working a 100 % and fire ignition will happen with a certain probability. Secondly, most houses don't have sprinklers or enclosed living rooms, including closed doors, to contain the fire in the living room. Therefore, Barrier 2 is not there and fire will grow with certainty. Thirdly, most houses don't have smoke control system to prevent the smoke from spreading to the whole house, including any egress paths such as stairs. Therefore, Barrier 3 is not there and smoke will spread with certainty. Fourthly, houses usually have smoke alarms to give early warnings. However, they are only effective if they work and work early so occupants can have enough time to escape. But fires involving upholstered furniture can be very fast, developing into flashover fires in just minutes. That is why the time factor is important in fire risk assessment. If the occupants can escape before the fire develops into a flashover fire, then Barrier 4 is there. Otherwise, Barrier 4 is not there and the occupants are trapped in the house. Their safety depends on the response of the fire department. If the fire department can respond quickly, then Barrier 5 is there. Otherwise, Barrier 5 is not there. In short, the only defence in house fires is Barrier 4 and 5, and they are only effective if they work and work early against fast fires. That is why house fires are deadly because they

Figure 2.3 House fires can be very rapid and therefore fatal (photo courtesy of Dr Joseph Su, reproduced by permission of the National Research Council Canada).

can be very fast and there are usually not enough barriers to protect the occupants. Figure 2.3 is a photo of a house fire experiment conducted by the National Research Council Canada.

For other types of occupancies, such as high-rise apartment and office buildings, there are usually more fire protection measures and therefore more barriers between the fire and the occupants. For example, in apartment buildings, there is usually compartmentation (each apartment unit is constructed as a fire compartment – Barrier 2) and sprinkler protection (Barrier 2) to contain the fire. Also, there are usually alarm systems (Barrier 4) to provide early warnings to the occupants and protected stairs (Barrier 4) to help the occupants to evacuate. With more barriers, fire risk assessment becomes more complex and is the subject of the subsequent chapters of this book.

2.3 Summary

In this chapter, the basic concepts of fire risk assessment were introduced. Fire risk assessment is the assessment of the risks to the people and property as a result of unwanted fires. A simple risk assessment considers the probability of the occurrence of a certain unwanted fire scenario and the consequence of that scenario. A comprehensive risk assessment considers all probable unwanted fire scenarios and their consequences.

A fire scenario is a set of fire events that are linked together by the success or failure of fire protection measures. There are basically five major hazardous events that must occur before a fire can cause harm to the occupants. They are: (1) fire ignition, (2) fire growth, (3) smoke spread, (4) failure of occupants to evacuate and (5) failure of fire department to respond. Each of these five hazardous events can be prevented from happening by fire protection measures, or barriers.

The probability of the fire scenario that can lead to harm to the occupants depends on the combined probability of failure of all fire protection measures, or barriers. The lower are the individual probabilities of failure of fire protection measures, the lower is the probability of the fire scenario that can lead to harm to the occupants. Fire risk assessment concerns not only the number of fire protection measures that are put in place, but also how reliable and effective these fire protection measures are.

The risk to the occupants depends not only on the probability of the fire scenario that can lead to harm to the occupants, but also the level of harm to the occupants as a result of the consequence of that scenario. The consequence of a fire scenario can be assessed by using time-dependent modelling of fire and smoke spread, occupant evacuation and fire department response.

2.4 Review Questions

2.4.1 If a fire starts in the living room within an apartment unit in an apartment building, how many fire barriers are there between the fire and the occupants in the unit?

2.4.2 If a fire starts in one apartment unit in an apartment building, how many fire barriers are there between the fire and the other occupants in the other apartment units on the same floor?

References

Custer, R.L.P. and Meacham, B.J. (1997) *Introduction to Performance-Based Fire Safety*, Society of Fire Protection Engineers and National Fire Protection Association, Quincy, MA, pp. 130–34.

Yung, D. and Benichou, N. (2003) *Concepts of Fire Risk Assessment*, Report No. NRCC-46393, National Research Council Canada, Ottawa, ON, pp. 1–4.

Yung, D. and Lougheed, G.D. (2001) *Fatal Fire Scenarios in Canadian Houses*, Internal Report No. 830, National Research Council Canada, Ottawa, ON, pp. 1–6.

3

Fire Risk Assessment Based on Past Fire Experience

3.1 Overview

In this chapter, we introduce a number of fire risk assessments, all of which are based on past fire experience. These fire risk assessments are only valid, however, in cases where the situation in the past and the present situation are similar or identical. For this purpose the controlling parameters governing the fire scenarios in both situations need to be the same. However, they are frequently not the same because of changes that take place over time – for instance, when new furnishing materials or new fire protection systems are introduced.

There are a number of controlling parameters including:

1. fire protection systems, such as sprinklers that control the development of a fire; or
2. alarm systems that expedite the evacuation of the occupants.

Controlling parameters also include a number of physical parameters, e.g.

1. the type and amount of combustibles governing the development of a fire; or
2. the number and length of the egress routes governing the required evacuation time.

Principles of Fire Risk Assessment in Buildings   D. Yung
© 2008 John Wiley & Sons, Ltd

Past experience can be a specific fire experience, such as the 2003 Station Club fire in Rhode Island, United States, or a general fire experience, such as those that are obtained from fire statistics. Examples of how to apply both experiences to the present are discussed in this chapter.

3.2 Based on Past Fire Experience

Fire risk assessments can be performed based on past fire experience. Such fire risk assessments, however, are valid only if the situation in the past and that to be assessed at the present are the same. This requires that the *controlling parameters* that govern the fire scenarios in both situations are the same. Often, they are not the same because of changes over time such as the introduction of new furnishing materials or new fire protection systems. Controlling parameters include fire protection systems, such as sprinklers that control the development of a fire or alarm systems that expedite the evacuation of the occupants. Controlling parameters also include physical parameters, such as the type and amount of combustibles that govern the development of a fire or the number and length of the egress routes that govern the required evacuation time. If these controlling parameters are not the same, then a fire risk assessment based on the past experience can be quite wrong.

The following two examples illustrate the importance of examining the controlling parameters to ensure that the fire scenarios that happened in the past and those that could happen in the present are similar before the fire experience from the past can be applied to the present. The first example is a deadly night club fire. A deadly fire often leads to an obligatory investigation of fire safety issues and the imposition of new safety regulations. As a result, the controlling parameters in the past and those at present are not the same. Past experience, therefore, may not apply. The second example is a house fire. House fires occur regularly with often tragic consequences. However, the number of deaths in a typical house fire, although tragic, is not at a level that would cause immediately major changes in regulations. As a result, the controlling parameters in the recent past and those at present may be the same. Recent past experience, therefore, may still apply. Long-time past experience, however, may not apply because of the changes over time such as the introduction of new furnishing materials or new fire protection systems.

3.2.1 Night Club Fire Scenario

We will look at a *night club fire* to examine whether fire experience from the past can be applied to a similar night club fire in the present. For this exercise, we will look at a night club fire with a simple and well-defined fire scenario. We will look at the February 2003 Station Club fire in Rhode Island, United States which killed around one hundred people. This fire started at the back of the stage while the band was playing. The fire was recorded by video because the show was being video recorded at the time (CNN News, 2003). The fire was also analysed by the National Institute of Science and Technology using both experimental and computer simulations (Madrzykowski, Bryner and Kerber, 2006).

Over the years, there have been many deadly night club fires in the world, but not many have simple and well-defined fire scenarios. For example, one of the deadliest night club fires in the United States is the 1942 Cocoanut Grove fire in Boston that killed 492 people. The location of the fire origin and the cause of that fast fire spread is still being analysed today after so many years (Beller and Sapochetti, 2000).

The Station Club was a small wooden building with a capacity for 300 people. There were four exits, including the main entrance, but no sprinklers. Fire started when fireworks used by the rock band to start the show ignited the combustible material on stage and the fire spread quickly. Most people died while trying to leave through the front entrance.

We will examine the controlling parameters of that Station Club fire and see whether those controlling parameters are common in other similar night clubs. If these controlling parameters remain the same, then the experience from the Station Club fire can be applied to the other night clubs. If they are not the same, then that experience cannot be applied to the other night clubs. We will first go through the five fire barriers which were discussed in Chapter 2.

1. Barrier 1 is a barrier to prevent a fire from starting. The rock band used fireworks on stage with plenty of easily combustible material around. The chance of starting a fire was very high. Barrier 1, therefore, was not there.
2. Barrier 2 is a barrier to contain the fire from spreading. The small night club was basically a large dance hall with no compartmentation to isolate the fire. The club also had no sprinklers to suppress the fire. Barrier 2, therefore, was not there.

3. Barrier 3 is a barrier to control the smoke from spreading. Since the small night club was just a large open dance hall, with no smoke control systems to either extract the smoke or to contain the smoke, smoke spread readily to the whole hall. Barrier 3, therefore, was not there.

4. Barrier 4 is a barrier to provide early warnings to the occupants and to safeguard the egress routes for evacuation. The rock band was playing at the time when the fire started, and was using fireworks for special effects, it would have been difficult for the patrons to notice a fire, or hear any instructions from anyone on how to evacuate. The patrons would not have been able to evacuate early, nor would they have been aware of all the exits that were available to them. Most left by the way they came in, therefore, using only one exit. Barrier 4, therefore, was not there.

5. Barrier 5 is to notify the fire department early so they can respond early. With nothing to control the fire, the fast fire would have engulfed the whole place quickly, probably in just minutes. It would have been difficult for the fire department to respond fast enough to control the fire and rescue the people. Barrier 5, therefore, was not there.

The Station Club, basically, had no fire barriers between the fire and the occupants. Fire barriers reduce the probabilities of fire scenarios that would lead to harm. When there were no barriers, the probabilities of fire scenarios that would lead to harm were 100 %. That explains why there were so many fatalities.

While fire barriers reduce the probabilities of fire scenarios that would lead to harm, the level of harm is governed by the physical parameters. We will look at the *physical parameters* in the Station Club fire that governed the two important time-dependent events: the speed of the fire development and the speed of occupant evacuation. The explanations below show that the physical parameters would make the speed of the fire development very fast and the speed of occupant evacuation very slow. That would explain why there were so many casualties.

1. Fire development was fast because the combustible on stage was the type of material that would burn quickly, such as foam and plastic. The dance hall was not large and a fast fire would fill up the space with smoke in just minutes.

2. Occupant evacuation was not fast because the patrons would not have noticed the fire early and would have been affected by the dense smoke. Also, most patrons apparently tried to leave via the main entrance door rather than using all the exits.

Let's examine the question whether we can apply the Station Club fire experience to other night clubs. The answer, as discussed before, is yes if the controlling parameters are the same; and no if the controlling parameters are not the same. To have the same controlling parameters in other night clubs requires that the bands in the other clubs would still be using fireworks on stage, that there would be similar fast burning material on stage, that there would be no fire protection systems in place, and that the people would still be using only one exit to leave. If these controlling parameters are not the same, then the risk is different. Often, as a result of a major fire incident, new fire safety regulations are imposed and night clubs with these new fire safety measures would have lower fire risks. For example, if fireworks are banned on stage, or the use of any pyrotechnics must follow a strict safety guideline, or sprinklers are required, or exits are required to be clearly marked and that the people are given prior instructions to use them before the show starts, the risk would be a lot lower.

3.2.2 Single House Fire Scenario

Fires in houses are different than the night club fire scenario that was discussed in the previous section. The previous night club fire scenario was a very specific one: the fire started on stage while the band was playing and the people were there watching. Therefore, the location of the origin of the fire and that of the people were both known. In a *house fire*, the fire could start anywhere in the house and the occupants could also be anywhere in the house. For example, a fire could start in the living room while the occupants are asleep in their bedrooms. Or, a fire could start in the kitchen while the occupants are nearby in the living room. Therefore, there could be many combinations of the origin of the fire and where the people are. The fire scenarios in a house need to be examined carefully before the fire experience from the past can be applied to the present. This is especially true when new building materials are being introduced from time to time.

We will go through the controlling parameters that govern fire scenarios to examine whether fire scenarios in the past and those that could happen in the present are the same. We will also discuss how fire experience can be grouped by the various controlling parameters to allow the fire scenarios that happened in the past and those that could happen in the present to be matched. For example, as will be discussed in this section, past experience can be grouped by the location of the origin of the fire, or by the presence and absence of certain type of fire protection system and so on. We will use this example to show why matching the controlling parameters would mean matching the fire scenarios.

We will first look at the fire protection systems or fire barriers in houses. We already had a discussion in Chapter 2 that there are usually not too much of a fire protection in a house. The probability of fatal fire scenarios, therefore, is relatively high when compared with other buildings. The usual protection in a house is the smoke alarm, which was introduced in the 1970s. Smoke alarms, however, work only if they are installed (some older houses may still not have them) and properly maintained (batteries are not removed and replaced regularly). Mandatory sprinklers have been introduced more recently in houses in some parts of the world. Because the cost of installing sprinklers in a house is relatively high, when compared with the cost of installing smoke alarms, there have been continuous debates on whether mandatory sprinklers are cost effective or not for houses. Economics aside, sprinklers do provide an additional protection in a house.

While smoke alarms help occupants to evacuate earlier (fire Barrier 4 in Chapter 2), sprinklers help control the fire from spreading (fire Barrier 2). If we group the experience based on the presence or absence of these fire protection systems, then we are matching the fire scenarios that happened in the past and those that could happen in the present (see discussions on fire scenarios in Chapter 2). This still requires that the other physical parameters that control the speed of fire development and those that control the speed of occupant evacuation are matched properly, which will be discussed next.

Physical parameters are those that control, for each fire scenario, the time-dependent development of the fire and the evacuation of the occupants. They include the architectural layout of the house and the combustibles in the house. The location of the origin of a fire is a random event, which will be discussed later in Chapter 7. Once a fire has started, the physical parameters control its development and also the evacuation of the occupants.

The location of the origin of a fire suggests the type of combustibles that is involved. For example, if the location of the origin of the fire is in the living area, the combustibles are probably the typical furniture in a living area. If we group the past experience by the location of the origin of a fire, then there is a good chance that the burning objects involved in the past and those at the present are similar. Consequently, the fire scenarios in the past and those in the present are probably similar. Past experience can therefore be applied to the present. However, as mentioned earlier, there is this caution that the fire experience in houses from a long time ago can still be different because of the introduction of new furnishing materials over the years.

The next section will discuss how fire statistics are grouped by occupancy type, by location of fire origin, and by fire protection systems so that they can be used for fire risk assessment.

3.3 Based on Fire Incident Data

Fire loss data are usually collected by many countries to obtain information on their country's fire losses and to help devise regulations to manage these loses. The loss information of a fire incident is usually collected by the responding firefighters, or by special fire investigators who are assigned to the case, and is usually recorded on standard fire incident report forms. The forms are then usually submitted to a department responsible for statistical analysis, which collects the data, analyses the data and produce the various statistical reports. These statistical reports provide valuable information for the assessment of future fire risks.

How good the fire statistics are depends on how good the fire incident data are. It depends on a number of factors. For example, it depends on the design of the fire incident report forms whether they capture all the required information, whether the required information is easily understood, and whether the required information is obtainable. It also depends on how accurate the required information is entered into the forms by the firefighters or the special investigators. Furthermore, it depends on the sample size of the total number of incidents that are collected. The larger the sample size, the more meaningful is the statistical analysis. A small country with a small population may not have enough fire incidents to provide the required sample size for a meaningful statistical analysis. Those with a large population may not have all the fire incidents reported to the data centre. For example, in the

United States, only one-third of the fire incidents that are collected by the fire departments are reported to the National Fire Incidents Reporting System (NFIRS). The sample size is still large enough that it can be scaled up to represent the whole country (Hall and Harwood, 1989).

3.3.1 International Fire Statistics

We have mentioned that fire statistics are routinely collected and produced by various countries. Comparisons of international fire statistics allow the various countries to look at their fire situation in relation to others and see whether they have improved or deteriorated. Such comparisons are produced annually by the World Fire Statistics Centre of the Geneva Association for the insurance companies. The 2005 comparisons (The Geneva Association, 2005), based on fire deaths during 2000–2002, are plotted here in Figure 3.1. It should be noted that the fire deaths for some of the countries are for years prior to 2000–2002. Also, the data for the United States includes the 2791 fire deaths resulting from the 9/11 World Trade Center collapse.

Figure 3.1 shows that during 2000–2002, the number of fire deaths per 100,000 persons in the majority of the listed countries is in the

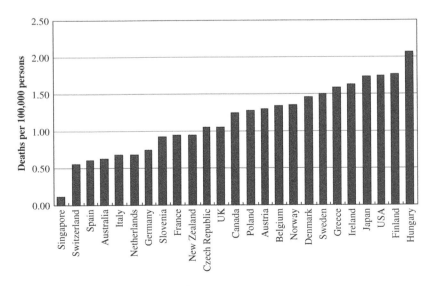

Figure 3.1 Comparison of fire deaths of various countries for years 2000–2002 (plotted from data in Table 4 of the World Fire Statistics Bulletin, No. 21, Oct 2005).

range of 0.6–1.8. Singapore is the exception with a low value of 0.12 fire deaths per 100,000 persons. Canada is in the middle of the pack, at 1.25. Hungary is at the high end, at 2.06.

3.3.2 National Fire Statistics

Most countries collect fire statistics to provide information on fire losses and to devise regulations to manage these losses. Statistical information includes both general information on the fire situation in the country and detailed information that can be used for fire risk assessment. For example, the Canadian Council of Fire Marshals and Fire Commissioners produce annual reports on fire losses, fire deaths and fire injuries as shown in Table 3.1. Detailed fire statistics usually stay with the provincial fire marshal's offices which will be discussed in the next section.

Table 3.1 shows not only the general information on fire losses, fire deaths and fire injuries, but also the trend over a 10-year period. The trend provides information on whether the fire situation in the country is improving or deteriorating. For example, the table shows that the number of fires dropped during the 10-year period, from 66,000 per year to 55,300 per year, and the per capita fire deaths also declined from 1.41 deaths per 100,000 persons to 1.09. The trend can also be used to compare with those of other countries to see whether the trend is unique in their country or is part of a general worldwide trend. For

Table 3.1 Canadian fire losses, fire deaths and fire injuries (source: Council of Canadian Fire Marshals and Fire Commissioners, 2001, Annual Report, Table 1).

Year	Estimated population (10^6)	Number of fires (10^3)	$ Loss (10^6)	Per capita $ loss	Fire deaths	Death rate (per 10^5 persons)	Injuries	Injuries Rate (per 10^5 persons)
1992	28.4	66.0	1241	43.8	401	1.41	3874	13.7
1993	28.7	65.9	1182	41.2	417	1.45	3463	12.1
1994	29.0	66.7	1152	39.7	377	1.30	3539	12.2
1995	29.4	64.3	1111	37.8	400	1.36	3551	12.1
1996	29.7	60.1	1163	39.2	374	1.26	3152	10.6
1997	30.0	56.3	1292	43.1	416	1.39	3149	10.5
1998	30.3	57.6	1176	38.8	337	1.11	2697	8.9
1999	30.5	55.2	1232	40.4	388	1.27	2287	7.5
2000	30.7	53.7	1185	38.6	327	1.06	2490	8.1
2001	31.1	55.3	1421	45.7	338	1.09	2310	7.4

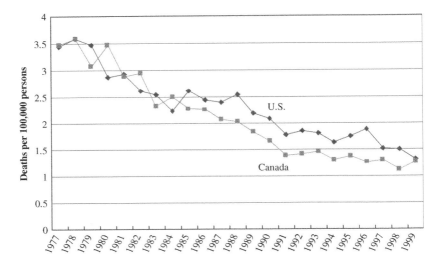

Figure 3.2   Fire deaths in the United States and Canada follow a similar decline (plotted from data in Hall, 2003).

example, Figure 3.2 is a comparison of fire deaths in the United States and Canada over a period of 23 years (Hall, 2003). Figure 3.2 shows that the fire deaths in the two countries follow a similar decline, from 3.5 deaths per 100,000 persons in 1977 to 1.3 in 1999. Such declines are the results of many factors, such as the introduction of smoke alarms in homes in the 1970s, and are the subjects of many statistical analyses. Changes in fire statistics over time show that the use of fire statistics for fire risk assessment requires careful considerations of what is changing. This will be discussed in more detail in the later chapters of this book on how to deal with these changes.

3.3.3 Use of Fire Statistics for Fire Risk Assessment

Fire loss information from fire incident reports is stored in databases that can be extracted for various statistical analyses (access to databases often requires special arrangements with collection agencies). For example, data can be extracted for certain type of occupancy, such as residential buildings. Within that occupancy type, further breakdown of the information can be obtained. For example, fire loss information can be obtained based on the area of fire origin, or source of ignition, or object first ignited and so on. Fire loss information can also be obtained based on the presence or absence of fire protection systems, such as smoke alarms or sprinklers. Following this approach, one can extract statistical information for a specific set of controlling parameters. For example, one can extract statistical information on fires originating in the living area of a single family house, with or without alarms or sprinklers. This allows the results to be applicable to situations with similar controlling parameters.

An example of this is the fire statistics extracted for Canadian houses for the three-year period from 1995 to 1997 (Yung and Lougheed, 2001), which are reproduced here in Figure 3.3. This figure shows both the percentage of house fires originating in different areas in a house and the associated fatality rates (deaths per 1000 fires originating in these different areas). It should be noted that this figure shows only those areas in a house with high fatality fires, and not all the fires in a house. The fatality rates help to show the severity of fires originating in different areas in a house. For example, the consequence of fires originating in the living area is 43.2 deaths per 1000 fires; whereas the consequence of those originating in the cooking area is 9.8 deaths per 1000 fires. The expected risk to life, however, depends not only on the severity of these

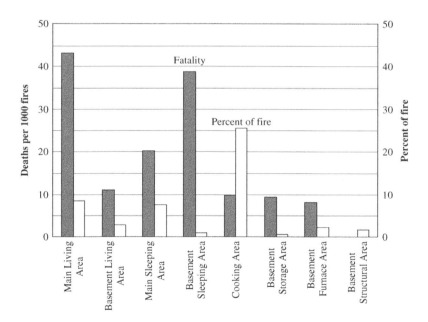

Figure 3.3 Fatality rates for fires originating in different areas in Canadian houses (deaths per 1000 fires) and the associated percentage of house fires originating in these areas, based on Ontario fire statistics 1995–1997 (from Yung and Lougheed, 2001, reproduced by permission of the National Research Council Canada).

fires, but also on the frequency of occurrence of these fires in different areas in a house.

The frequency of fire occurrence can also be obtained from fire statistics. As an example to show how it can be done, the 1996 Canadian census data and the corresponding 1996 Canadian fire statistics are referenced here. In the 1996 Canadian census data (Statistics Canada, 1996), the total number of houses under the category of 'single detached houses' and under the category of 'semi-detached houses, row and duplex houses, mobile homes and apartments in buildings under 5-storeys' was 9,840,580. In the 1996 Canadian fire statistics (Council of Canadian Fire Marshals and Fire Commissioners, 1996), the number of house fires under the category of 'one and two family dwellings' and under the category of 'rooming and mobile homes' was 17,232. It should be noted that fire statistics and census survey don't necessarily use the same categories to collect their information. To obtain the required information, one can only try to find the best match of the available categories of different databases. The frequency of fire occurrence can be obtained by dividing the number of fires by the number of houses, which is equal to 1.75×10^{-3} fires/house/year.

Similarly, the same fire statistics show that the total number of fire deaths in those houses in that year was 191. The risk of dying in a house fire, therefore, was 191 deaths divided by 9,840,580 houses, or 1.94×10^{-5} deaths/house/year.

The expected risk to life for fires originating in different areas in a Canadian house can be obtained using the above information on the frequency of fire occurrence in houses and the information in Figure 3.3 on the percentage of fires originating in different areas in a house and the fatality rates of fires originating in different areas. The results are listed in Table 3.2, in the order of the expected risk to life values for fires originating in different areas in a house. As in Figure 3.3, this table shows only those areas in a house with high fatality fires, not all the fires in a house. The results show that the highest expected risk to life is from fires originating in the living area, follow by fires originating in the cooking area, and then by those originating in the sleeping area. The cumulative risk to life from all these fires originating in different areas in a house, accounting for 50 % of house fires, is 1.51×10^{-5} deaths/house/year.

If all fires in a house are included, the cumulative risk to life, as was shown earlier, is 1.94×10^{-5} deaths/house/year. This number can be interpreted in two ways. It could mean the risk of dying in one's own house is 1.94 deaths in 100,000 years, which is a small number.

Table 3.2   Expected risk to life for fires originating in different areas in Canadian houses in 1996, with the probability of fire occurrence at 1.75×10^{-3} fires/house/year. This table shows only those areas in a house with high fatality fires, and not all the fires in a house.

Area of fire origin	% of house fires	Severity (deaths/fire)	Risk (deaths/house/year)
Main living	8.5 %	43.2×10^{-3}	6.43×10^{-6}
Cooking	25.6 %	9.8×10^{-3}	4.39×10^{-6}
Main sleeping	7.5 %	20.2×10^{-3}	2.65×10^{-6}
Basement sleeping	1.0 %	38.7×10^{-3}	6.78×10^{-7}
Basement living	2.8 %	11.0×10^{-3}	5.39×10^{-7}
Basement furnace	2.3 %	8.1×10^{-3}	3.26×10^{-7}
Basement storage	0.7 %	9.3×10^{-3}	1.14×10^{-7}
Basement structural	1.6 %	0.0	0.0
Total	50.0%		1.51×10^{-5}

However, it could also mean, with approximately 10 million houses in Canada, a risk of losing 194 of its citizens in a year, which is not a small number.

The above assessment of the risk to life values is, obviously, only good for a Canadian house in 1996. With the introduction of new furnishing materials and fire protection measures over the years, the risk values change with time. Also, not all houses are the same. The assessed risk values are, therefore, average values for Canadian houses. The objective here, however, is not to work out the expected risk to life values in Canadian houses. The objective is mainly to show how risk values can be assessed using fire statistics. If fire statistics are not available, or not current, then more fundamental approach using mathematical modelling of fire development and occupant evacuation is needed. This will be discussed in later chapters in this book.

3.4 Summary

Fire risk assessments based on past fire experience can be performed. Such fire risk assessments, however, are valid only if the situation in the past and that to be assessed at the present are the same. This requires that the controlling parameters that govern the fire scenarios in both situations are the same. Often, they are not the same because of changes over time such as the introduction of new furnishing materials or new fire protection systems.

Controlling parameters include fire protection systems, such as sprinklers that control the development of a fire or alarm systems that expedite the evacuation of the occupants. Controlling parameters also include physical parameters, such as the type and amount of combustibles that govern the development of a fire or the number and length of the egress routes that govern the required evacuation time. Two examples were used to illustrate the importance of examining these controlling parameters to ensure that the fire scenarios that happened in the past and those that could happen in the present are similar before the fire experience from the past can be applied to the present. The two examples were the 2003 Station Club fire in Rhode Island, United States and a typical single house fire.

Fire statistics from the past can also be applied to the present for fire risk assessment. They should be applied to situations with similar controlling parameters. For example, statistics based on fires originating in the living area of a single family house, with or without alarms or sprinklers, can be applied to similar situations if all other controlling parameters are the same. Often they are not the same because of the introduction of new furnishing materials and fire protection measures over the years.

3.5 Review Questions

3.5.1 If sprinklers are installed in a night club with a reliability of suppressing 95 % of the fires, how much of the risk is reduced? Review the barriers in Chapter 2 and also the referenced article (Madrzykowski *et al.*, 2006).

3.5.2 If fire prevention in a household is successful in lowering the fire occurrence in a living area to 85 % of it normal value, how much lower is the expected risk to life for fires originating in the living area in a Canadian house?

References

Beller, D. and Sapochetti, J. (2000) Searching for answers to the Cocoanut Grove. *NFPA Journal*, May/June, 84–92.

CNN News (2003) At Least 96 Killed in Night Club Inferno, 21 Feb 2003, www.cnn.com/2003/US/Northeast/02/21/deadly.nightclub.fire/

Council of Canadian Fire Marshals and Fire Commissioners (1996) Fire Losses in Canada, Annual Report, Table 3b, http://www.ccfmfc.ca/stats/en/report_e_96.pdf

Council of Canadian Fire Marshals and Fire Commissioners (2001) Fire Losses in Canada, Annual Report, Table 1, http://www.ccfmfc.ca/stats/en/report_e_01.pdf

The Geneva Association (2005) World Fire Statistics Bulletin, No. 21, Table 4, October 2005, www.genevaassociation.org

Hall, J.R., Jr. (2003) Fire in the U.S. and Canada, NFPA Report, Quincy, MA, April 2003, p. 3.

Hall, J.R., Jr. and Harwood, B. (1989) The national estimates approach to U.S. Fire statistics. *Fire Technology*, **25** (2), 99–113.

Madrzykowski, D., Bryner, N. and Kerber, S.I. (2006), The NIST Station night club fire investigation: physical simulation of the fire. *Fire Protection Engineering Magazine*, Summer 31, 35–46, http://www.fpemag.com/archives/article.asp?issue_id=37&i=245

Statistics Canada (1996) Census Canadian Private Households by Size, http://www.statcan.ca/english/census96/oct14/hou.pdf

Yung, D. and Lougheed, G.D. (2001) Fatal Fire Scenarios in Canadian Houses, IRC Internal Report No. 830, National Research Council Canada, Ottawa, ON, October 2001, Figure 1.

4

Qualitative Fire Risk Assessment

4.1 Overview

In this chapter, qualitative fire risk assessment is introduced. *Qualitative fire risk assessment* is based on subjective judgment of not only the probability of a fire hazard or fire scenario occurring, but also the consequence of such a fire hazard or fire scenario. The term *fire hazard* generally describes any fire situation which is dangerous and which may have potentially serious consequences; whereas the term *fire scenario* was defined in Chapter 2 previously as a sequence of fire events that are linked together by whether the fire protection measures succeeded or failed. Qualitative fire risk assessment is usually employed in order to obtain a quick assessment of the potential fire risks in a building and to consider various fire protection measures to minimize these risks.

In general qualitative fire risk assessments may be performed in two ways:

1. a checklist is used to go through the potential fire hazards, the fire protection measures to be considered and the subjective assessment of their fire risks;
2. an event tree is used to go through the potential fire scenarios and the fire protection measures to be considered and the subjective assessment of their fire risks.

The outcome in both cases, is a list of potential fire hazards, or fire scenarios, the fire protection measures to be considered and their assessed

Principles of Fire Risk Assessment in Buildings   D. Yung
© 2008 John Wiley & Sons, Ltd

fire risks. In this context assessed risks are described in qualitative rather than quantitative terms.

4.2 Risk Matrix

Fire risk is measured, as described in Chapter 2, by the product of the probability of occurrence of a fire scenario and the consequence of that scenario. In qualitative fire risk assessments, there are no numerical values for the probability or consequence that can be used to obtain the product. Instead, the product is assessed using a simple two-dimensional *risk matrix*, with one axis representing the level of the probability of occurrence and the other representing the severity of the consequence. The degree of risk is assessed based on how high the probability is and how severe the consequence is. An example of a risk matrix is shown in Figure 4.1 (drawn here from Table E3, Standards Association of Australia, 1999, 'Risk Management'). In this risk matrix, the value of the probability is divided into five levels and the severity of the consequence is divided into five categories. The higher the probability and the higher the consequence in the matrix, the higher is the assessed risk (similar to the product of two values). For example, the combination of an

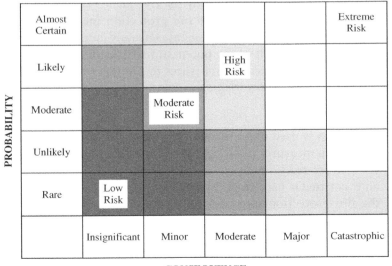

Figure 4.1   Risk matrix diagram where the degree of risk is assessed based on the level of the probability of occurrence and the severity of the consequence (drawn here from Table E3, Standards Australia AS/NZS4360, 1999, 'Risk Management').

'almost certain' probability and a 'catastrophic' consequence is assessed as an 'extreme' risk; whereas the combination of a 'rare' probability and an 'insignificant' consequence is assessed as a 'low' risk. In between these two extremes, the risk is assessed as either 'moderate' or 'high', depending on the combination of the probability and the consequence.

In qualitative fire risk assessments, as was described earlier, various terms are used to describe the values of the probability, the consequence and the assessed risk. It should be noted that there are no standards on how to name these terms. Usually, these terms are developed for specific applications. For example, the definitions of the terms used in Figure 4.1, shown in Table 4.1, were developed mainly for occupational health and safety risk assessments in Australia and New Zealand. Similar definitions

Table 4.1 Definitions of probability, consequence and risk levels used in Figure 4.1 (source: Tables E1-E3, Standards Association of Australia, 1999, 'Risk Management', developed mainly for occupational health and safety risk assessments).

Probability Level	Definition
Almost certain	Is expected to occur in most circumstances
Likely	Will probably occur in most circumstances
Moderate	Might occur at some time
Unlikely	Could occur at some time
Rare	May occur only in exceptional circumstances

Consequence Level	Definition
Catastrophic	Death, toxic release off-site with detrimental effect, huge financial loss
Major	Extensive injuries, loss of production capability, off-site release with no detrimental effects, major financial loss
Moderate	Medical treatment required, on-site release contained with outside assistance, high financial loss
Minor	First aid treatment, on-site release immediately contained, medium financial loss
Insignificant	No injuries, low financial loss

Risk Level	Definition
Extreme	Immediate action required
High	Senior management action required
Moderate	Management responsibility specified
Low	Managed by routine procedures

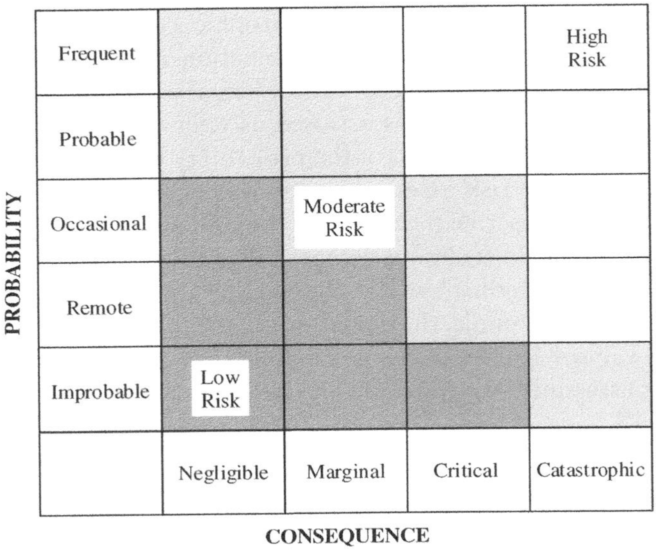

PROBABILITY		Negligible	Marginal	Critical	Catastrophic
Frequent					High Risk
Probable					
Occasional			Moderate Risk		
Remote					
Improbable		Low Risk			
		Negligible	Marginal	Critical	Catastrophic

CONSEQUENCE

Figure 4.2 Risk matrix diagram originally developed for military application (Reprinted with permission from NFPA 551 – 2007: Guide for the Evaluation of Fire Risk Assessment, Copyright © 2007, National Fire Protection Association, Quincy, MA, USA. This reprinted material is not the complete and official position of the NFPA on the reference subject, which is represented only by the standard in its entirety).

can be developed for other applications. Figure 4.2 shows a different risk matrix which was developed for military applications (reproduced here from Figure A.5.2.5, NFPA 551, 2007, 'Guide for the Evaluation of Fire Risk Assessments'), with definitions shown in Table 4.2 (source: Tables A.5.2.5(a) and (b), NFPA 551, 2007, 'Guide for the Evaluation of Fire Risk Assessments'). Another example is shown in Figure 4.3 (drawn differently from Figure 10-2, SFPE, 2000, 'Engineering Guide to Performance-Based Fire Protection Analysis and Design of Buildings'), with definitions shown in Table 4.3 (source: Tables 10-1 and 10-2, SFPE, 2000, 'Engineering Guide to Performance-Based Fire Protection Analysis and Design of Buildings').

4.3 Checklist Method

The *checklist method* (NFPA 551, 2007) employs the creation of a checklist of potential fire hazards and the consideration of fire protection

Table 4.2 Definitions of probability and consequence levels used in Figure 4.2 (source: Tables A.5.2.5(a) and (b), NFPA 551, 2007, 'Guide for the Evaluation of Fire Risk Assessments').

Probability Level	Definition
Frequent	Likely to occur frequently
Probable	Will occur several times during system life
Occasional	Unlikely to occur in a given system operation
Remote	So improbable, may be assumed this hazard will not be experienced
Improbable	Probability of occurrence not distinguishable from zero.

Consequence Level	Definition
Catastrophic	The fire will produce death or multiple deaths or injuries. The impact on operations will be disastrous, resulting in long-term or permanent closing. The facility would cease to operate immediately after the fire occurred.
Critical	Personal injury and possibly deaths may be involved. The loss will have a high impact on the facility, which may have to suspend operations. Significant monetary investments may be necessary to restore to full operations.
Marginal	Minor injury may be involved. The loss will have impact on the facility, which may have to suspend some operations briefly. Some monetary investments may be necessary to restore the facility to full operations.
Negligible	The impact of loss will be so minor that it would have no discernible effect on the facility or its operations.

measures, either in place or to be added, to arrive at a subjective judgment of the fire risks. The creation of a checklist of potential fire hazards allows a systematic check of potential fire hazards that are in place. The listing of fire protection measures alongside with the potential fire hazards allows a quick check of any safety deficiencies and any need to provide additional fire protection measures to minimize the risk. The checklist method, therefore, is an enumeration of potential fire hazards, fire protection measures, either in place or to be added, and the subjective judgment of the residual fire risks. It is used to identify any deficiencies and any corrective measures needed to minimize the fire risks. It does not include, however, the consideration of the logical development of fire events, which will be discussed in Section 4.4 using an event tree.

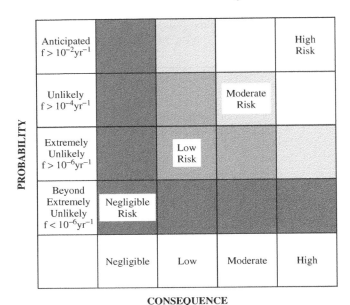

Figure 4.3   Another example of a risk matrix diagram (drawn differently from Figure 10.2, SFPE, 2000, 'Engineering Guide to Performance-Based Fire Protection Analysis and Design of Buildings').

An example of a checklist method employing qualitative fire risk assessment is shown in Table 4.4. This example looks at a potential fire hazard in the living room of a house and the consideration of a number of additional fire protection measures to minimize the risk. Obviously, there could be many fire hazards in a house. A complete fire risk assessment would involve the identification of all potential fire hazards and the consideration of various fire protection measures to minimize the risk.

A typical house usually has some fire protection measures, such as smoke alarms. Additional fire protection measures would lower the risk further. This example considers six different combinations of three additional fire protection measures. The three additional fire protection measures are: (1) no smoking material (such as cigarettes) in the living room, (2) sprinklers and (3) regular evacuation drills. Each of the three fire protection measures has an impact on either the probability of fire occurrence or the consequence of a fire occurrence. For example, the measure of 'no smoking material in the living room' would have an impact on lowering the probability of fire occurrence. The measures of 'sprinklers' and 'regular evacuation drills' would have an impact on lowering the

Table 4.3 Definitions of probability and consequence levels used in Figure 4.3 (source: Tables 10-1 and 10-2, SFPE, 2000, 'Engineering Guide to Performance-Based Fire Protection Analysis and Design of Buildings').

Probability Level	Description	Frequency (median time to event)
Anticipated	Incidents that might occur several times during the lifetime of the building.	$>10^{-2}$/yr (<100 yr)
Unlikely	Events that are not anticipated to occur during the lifetime of the facility.	10^{-4}/yr $< f < 10^{-2}$/yr ($100–10\,000$ yr)
Extremely unlikely	Events that will probably not occur during the life cycle of the building.	10^{-6}/yr $< f < 10^{-4}$/yr ($10\,000–1\,000\,000$ yr)
Beyond extremely unlikely	All other accidents	$<10^{-6}$/yr ($>1\,000\,000$ yr)

Consequence Level	Impact on populace	Impact on property/operations
High	Sudden fatalities, acute injuries, immediately life threatening situations, permanent disabilities	Damage > \$X million, Building destroyed, surrounding property damaged
Moderate	Serious injuries, permanent disabilities, hospitalization required	\$Y < damage < \$X million Major equipment destroyed, minor impact on surroundings
Low	Minor injuries, no permanent disabilities, no hospitalization	Damage < \$Y million, Reparable damage to building, significant operational downtime, no impact on surroundings
Negligible	Negligible injuries	Minor repairs to building required, minimal operational downtime

consequence of a fire occurrence by suppressing or controlling the fire or by allowing the occupants to evacuate more quickly.

It should be noted that Table 4.4 is only an example of a checklist method, not necessarily a standard method in fire risk assessment. It is used here to show that, in qualitative fire risk assessment, the definitions of the terms for the various levels of probability, consequence and risk need to be developed first by the stakeholders before the fire risk assessment can be carried out. The definitions below are assumed by the author as an example; there are no standard definitions.

Table 4.4 Qualitative fire risk assessment of a potential fire hazard in the living room of a house using the checklist method.

No	Fire Hazard	Existing Fire Protection	Inherent fire risk			Additional Fire Protection	Residual fire risk		
			P	C	Risk		P	C	Risk
1	Fire in living room	Smoke Alarms	Unlikely	Critical	Moderate	No smoking material in living room	Extremely Unlikely	Critical	Low
2	Fire in living room	Smoke alarms	Unlikely	Critical	Moderate	Sprinklers	Unlikely	Marginal	Low
3	Fire in living room	Smoke alarms	Unlikely	Critical	Moderate	Evacuation drills	Unlikely	Marginal	Low
4	Fire in living room	Smoke alarms	Unlikely	Critical	Moderate	No smoking material in living room + sprinklers	Extremely unlikely	Marginal	Low
5	Fire in living room	Smoke alarms	Unlikely	Critical	Moderate	No smoking material in living room + evacuation drills	Extremely unlikely	Marginal	Low
6	Fire in living room	Smoke alarms	Unlikely	Critical	Moderate	Sprinklers + evacuation drills	Unlikely	Negligible	Negligible

Note: In this table, P is the probability of the fire hazard and C is the consequence of the fire hazard. Only the control measure of no smoking material in living room affects the consequence. Other fire control measures affect the consequence.

In Table 4.4, the value of the probability is divided into the same four levels as in Table 4.3. The definitions of the probability levels are also similar to those in Table 4.3 and are shown in Table 4.5. If we assume that the living room fire is 'not anticipated to occur during the lifetime of the house', then the probability level is rated as 'unlikely', as shown in Table 4.5. If fire prevention can be established to allow 'no smoking material in the living room', then the fire 'will probably not occur during the life cycle of the house' and the probability level is lowered to 'extremely unlikely', as shown in Table 4.5.

In Table 4.4, the severity of the consequence is divided into the same four levels as in Table 4.2. The definitions of the severity levels are modified from those in Table 4.2 and are shown in Table 4.5 which shows that the severity of the consequence is based on what additional fire protection measures are in place. For example, if there are 'no additional sprinklers or evacuation drills', the consequence will be 'some occupants escape with some deaths' and the severity is given a 'critical' level. If there are 'either sprinklers or evacuation drills', the consequence will be 'all occupants escape with some injuries' and the severity is given a 'marginal' level. If there are 'both sprinklers and evacuation drills', the consequence will be 'all occupants escape with no injuries' and the severity is given a 'negligible' level.

Table 4.5   Definitions of probability and consequence levels used in Table 4.4.

Probability Level	Definition
Anticipated	Might occur several times during the lifetime of the house
Unlikely	Not anticipated to occur during the lifetime of the house
Extremely unlikely	Will probably not occur during the life cycle of the house (no smoking material)
Beyond Extremely unlikely	Less than extremely unlikely to occur

Consequence Level	Definition
Catastrophic	Many deaths
Critical	Some occupants escape with some deaths (no sprinklers or evacuation drills)
Marginal	All occupants escape with some injuries (with either sprinklers or evacuation drills)
Negligible	All occupants escape with no injuries (with both sprinklers and evacuation drills)

After the probability and consequence levels have been defined, the *inherent fire risks* and the *residual fire risks* in Table 4.4 can be obtained using a risk matrix. Inherent risks are risks before any additional fire protection measures are considered; whereas residual risks are the reduced risks if additional fire protection measures are put in place. For this example, the risk matrix is assumed to be the same as shown in Figure 4.3, but with the names of the severity of the consequence changed to those as shown in Table 4.5. That is, 'High' is changed to 'Catastrophic'; 'Moderate' to 'Critical'; 'Low' to 'Marginal' and 'Negligible' stays the same as 'Negligible'. In the table, P is the probability of the fire hazard and C is the consequence of the fire hazard. Only the fire prevention measure of no smoking material in the living room affects the probability of occurrence of fires in the living room. All other fire control measures affect the consequence.

The results in Table 4.4 show that the inherent risk is 'moderate', but with additional fire protection measures, the risk can be lowered to, 'low', or 'negligible'.

4.4 Event-Tree Method

An *event tree* is another way to identify potential fire hazards, judge their probabilities and consequences and arrive at risk ratings. Different from the checklist method, an event tree shows more than a list of potential fire hazards and fire protection measures for the judgment of the probabilities, consequences and eventually the risk ratings. The event-tree method constructs an event-tree subsequent to the initiation of a fire hazard, as described in Chapter 2, which provides more information for the judgment of probability, consequence and risk rating. An example for a fire hazard in an assumed apartment building is shown in Figure 4.4.

In Figure 4.4, the branching to different events depends on the success or failure of the fire protection measures in place. This example looks at one fire hazard in an assumed apartment building and the consideration of a number of additional fire protection measures to minimize the risk. The same event tree can be constructed for more hazards and more fire protection measures. A complete fire risk assessment would involve the identification of all potential fire hazards and the consideration of various fire protection measures to minimize the risk.

A typical apartment building usually has some fire protection measures, such as fire resistant construction and fire alarms. Additional fire protection measures would lower the risk further. This example

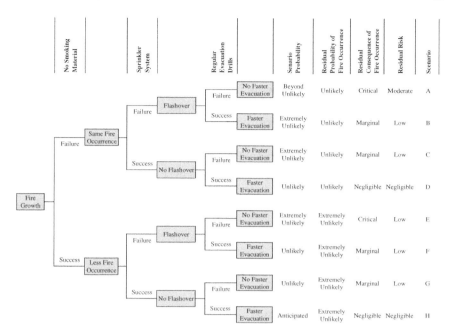

Figure 4.4 An example of an event tree for the judgement of probability, consequence and risk rating for the various fire scenarios resulting from a fire occurrence in an apartment building.

considers the same six different combinations of three additional fire protection measures as were used in the checklist method in Section 4.3. The three additional fire protection measures are: (1) no smoking material (such as cigarettes) in the apartments, (2) sprinklers, and (3) regular evacuation drills. Each of the three fire protection measures has an impact on either the probability of fire occurrence or the consequence of a fire occurrence. For example, the measure of 'no smoking material in the apartment' would have an impact on lowering the probability of fire occurrence; whereas the measures of 'sprinklers' and 'regular evacuation drills' would have an impact on lowering the consequence of a fire occurrence by suppressing or controlling the fire or by allowing the occupants to evacuate more quickly.

As is in the discussion of the checklist method, the event tree in Figure 4.4 is only an example to show how an event tree can be used for qualitative fire risk assessment. The description of each event is assumed by the author for this example only. Other applications may have different descriptions. The descriptions allow more transparent discussions and agreements among stakeholders.

In an event tree, each fire scenario has a probability value depending on the success or failure of the fire protection measures associated with that scenario. For this example, the level of probability is again divided into the same four levels as in Table 4.5. The definitions of the probability levels are different than those in Table 4.5 and are shown in Table 4.6. The definitions are assumed to be based on the number of successes and failures of the fire protection measures associated with the scenario, with a further assumption that the probability of failure of each fire protection measure is a much smaller value than that of the probability of success. For example, in Table 4.6, the scenario probability is given an 'anticipated' level if the scenario has a 'zero failure' of fire protection measures, and a 'beyond unlikely' level if the scenario has a 'three failures' of fire protection measures. The scenario probability for each scenario is shown in Figure 4.4. For example, Scenario A, with a 'three failures' of fire protection measures, is given a scenario probability of 'beyond unlikely'.

Table 4.6   Definitions of probability and consequence levels used in Figure 4.4.

Scenario probability Level	Definition
Anticipated	Zero failure of fire protections measures
unlikely	One failure of fire protections measures
Extremely unlikely	Two failures of fire protections measures
Beyond unlikely	Three failures of fire protections measures

Residual probability of fire occurrence Level	Definition
Anticipated	Might occur several times during the lifetime of the building
unlikely	Not anticipated to occur during the lifetime of the building
Extremely unlikely	Will probably not occur during the life cycle of the building (no smoking material)
Beyond unlikely	Less than extremely unlikely to occur

Residual consequence of fire occurrence Level	Definition
Catastrophic	Many deaths
Critical	Some occupants escape with some deaths (failures of both sprinkler system and evacuation drills)
Marginal	All occupants escape with some injuries (failure of either sprinkler system or evacuation drills)
Negligible	All occupants escape with no injuries (no failure of either sprinkler system or evacuation drills)

For this example, the value of the residual probability of fire occurrence for each fire scenario is also divided into the same four levels as in Table 4.5. The definitions of the residual probability levels are similar to those in Table 4.5 and are shown in Table 4.6. If we assume that the fire occurrence in an apartment 'is not anticipated to occur during the lifetime of the apartment', then the probability level is rated as 'unlikely', as shown in Table 4.6. If fire prevention can be established to allow 'no smoking material in the apartment', then the fire 'will probably not occur during the life cycle of the house' and the probability level is lowered to 'extremely unlikely', as shown in Table 4.6. The residual probability of fire occurrence for each scenario is shown in Figure 4.4. For example, Scenarios A, B, C, D, all with a failure of implementing 'no smoking material', have a residual probability of fire occurrence of 'unlikely', whereas Scenarios E, F, G, H, with a success of implementing 'no smoking material', have a residual probability of fire occurrence of 'extremely unlikely'.

The severity of the residual consequence for each fire scenario is also divided into the same four levels as in Table 4.5. The definitions of the severity levels are similar to those in Table 4.5 and are shown in Table 4.6. Table 4.6 shows that the severity of the residual consequence is based on how many of the additional fire protection measures fail. For example, if the sprinklers fail to operate and the regular evacuation drills fail to be implemented, the consequence will be 'some occupants escape with some deaths' and the severity is given a 'critical' level. If either the sprinklers fail to operate or the regular evacuation drills fail to be implemented, the consequence will be 'all occupants escape with some injuries' and the severity is given a 'marginal' level. If there is no failure of the sprinklers to operate and no failure of the evacuation drills to be implemented, the consequence will be 'all occupants escape with no injuries' and the severity is given a 'negligible' level. The residual consequence of fire occurrence for each scenario is shown in Figure 4.4. For example, Scenarios A, with a failure of both the 'sprinklers' and 'evacuation drills', has a residual consequence of 'critical', whereas Scenarios B, with a failure of only the 'sprinklers', has a residual consequence of 'marginal'.

Using the residual probability and residual consequence levels in each scenario, the residual risk under that scenario can be obtained using a risk matrix. For this example, the risk matrix is assumed to be the same as shown in Figure 4.3, but with the names of the severity of the consequence changed to those as shown in Table 4.6. That is, 'High' is changed to 'Catastrophic'; 'Moderate' to 'Critical'; 'Low' to 'Marginal' and 'Negligible' stays the same as 'Negligible'. Figure 4.4 shows the residual risks are reduced from 'moderate' to 'low' or 'negligible', depending on the success or failure of the additional fire protection measures. In Figure 4.4, each fire scenario has a scenario probability and a residual risk rating. For example, Scenario A has scenario probability of 'beyond unlikely' and a residual risk of 'moderate'. In qualitative risk assessment, it is not possible to combine all the scenarios to come up with a single residual risk because the ratings are not numerical. Combined rating can only be obtained using quantitative risk assessment which will be discussed in Chapter 5.

It should be emphasized again that the definitions in this example are not necessarily those that are in actual practice. They are assumed here by the author as an example. There are no standard definitions and the definitions need to be agreed upon by stakeholders before the risk assessment is carried out.

4.5 Summary

Qualitative fire risk assessment is an assessment based on subjective judgment of both the probability of occurrence of a fire hazard, or fire scenario, and the consequence of that fire hazard, or fire scenario. A *fire hazard* is a term generally used to describe a dangerous fire situation with potentially serious consequences; whereas a *fire scenario* is defined previously in Chapter 2 as a sequential set of fire events that are linked together by the success or failure of certain fire protection measures.

Qualitative fire risk assessment is employed usually as a quick assessment of the potential fire risks in a building and the consideration of various fire protection measures to minimize these risks. The assessment is non-quantitative and uses qualitative words such as 'likely' or 'unlikely' to describe probability, or 'major' or 'minor' to describe consequence. Judgment of both the probability of occurrence and the consequence of a fire hazard, or fire scenario, includes the consideration of how they are affected by various fire protection measures. Because there are no numerical values for both the probability and the consequence, the product of these two quantities is evaluated using a risk matrix.

There are in general two ways to conduct qualitative fire risk assessments: (1) use a checklist to go through the potential fire hazards, the fire protection measures to be considered, and the subjective assessment of their fire risks; (2) use an event tree to go through the potential fire scenarios and the fire protection measures to be considered and the subjective assessment of their fire risks. In both cases, the outcome is a list of potential fire hazards, or fire scenarios, the fire protection measures to be considered and their assessed fire risks. The assessed risks are described in qualitative, not quantitative, terms. Quantitative fire risk assessment is discussed in the next chapter.

4.6 Review Questions

4.6.1 Use the checklist method to do a qualitative fire risk assessment for a night club fire. Change 'no smoking material' to 'no fireworks on stage', and change 'evacuation drills' to 'prior evacuation instructions before show starts'. Make your assumptions for definitions for levels of probability, consequence and the risk matrix. Review the night club fire discussion in Chapter 3.

4.6.2 Use the event-tree method to do a qualitative fire risk assessment for a night club fire. Change 'no smoking material' to 'no fireworks on stage', and change 'evacuation drills' to 'prior evacuation instructions before show starts'. Make your assumptions for definitions for levels of probability, consequence and the risk matrix. Review the night club fire discussion in Chapter 3.

References

NFPA 551 (2007) *Guide for the Evaluation of Fire Risk Assessments*, National Fire Protection Association, Quincy, MA.

SFPE (2000) *Engineering Guide to Performance-Based Fire Protection Analysis and Design of Buildings*, National Fire Protection Association, Quincy, MA.

Standards Association of Australia (1999) Standards Australia and Standards New Zealand AS/NZS4360 Risk Management, Strathfield, NSW.

5

Quantitative Fire Risk Assessment

5.1 Overview

In this chapter, quantitative fire risk assessment is introduced. The term *quantitative fire risk assessment* refers to an assessment involving numerical quantifications not only of the probability a fire hazard, or fire scenario occurring, but also the consequences of that fire hazard or fire scenario. By multiplying the numerical values of probability and consequence each fire scenario is given a numerical fire risk value. By accumulating the sum of the risk values from all probable fire scenarios we can obtain an overall fire risk value. The overall fire risk value can be used for comparisons with those of alternative or code-compliant fire safety designs.

In general there are two ways to perform systematic quantitative fire risk assessments as follows:

1. by using a checklist to go through a list of potential fire hazards and the quantitative assessment of their fire risks;
2. by using an event tree to go through a set of potential fire scenarios and the quantitative assessment of their fire risks.

In both these methods, the values for the probability and consequence parameters are obtained from statistical data, if they are available, or from subjective judgment, if such data are not available.

5.2 Risk Indexing

Risk indexing involves the use of a set of well-defined risk parameters that have been developed for a specific application. The parameters can be both risk parameters (contributing to risk) and safety parameters (contributing to safety). The value of each parameter can be selected, based on its characteristics, from well-defined tables that have been developed by experts specifically for this application. The assessed values (index) can be used for comparison with those of mandatory requirements, or for comparison with those of alternative fire protection measures. In risk indexing methods, there are no separation of probability and consequence. Each parameter is given an assessed value and the summation of all these values are used for comparisons for compliance or equivalency.

One such representative risk indexing method is the one developed by NFPA (National Fire Protection Association) for health care facilities (NFPA 101A, 2004). In the NFPA method, worksheets are used to evaluate whether a facility can meet the basic safety requirements in four areas: (1) containment, (2) extinguishment, (3) people movement and (4) general safety. Table 5.1 is an example of what one of these worksheets looks like. Worksheet 4.7.7 has a list of 13 safety parameters which are to be evaluated under these four safety areas. The value for each of these 13 safety parameters is actually worked out in a separate worksheet. Their values are then entered into this Worksheet 4.7.7. The sum of all values in one column (one safety area) represents the evaluated total value for that safety area. For example, the sum of all values in the column for S_1 represents the evaluated total value for containment safety. The total value in each safety area is then compared with the required value for that safety area. The facility is considered safe if the evaluated total values meet the required values in all four areas. For more details of this method, consult the reference (NFPA 101A, 2004).

Other risk indexing methods are similar in concept, but with different sets of parameters and tables for different applications. They can be found in the SFPE (Society of Fire Protection Engineers) Handbook (Watts, 2002).

5.3 Checklist Method

As was discussed in Chapter 4, the *checklist method* employs the creation of a checklist of potential *fire hazards* and the consideration of fire protection measures, either in place or to be added, to arrive at an assessment of the fire risks. The creation of a checklist of potential

Table 5.1 NFPA safety evaluation worksheet (source: NFPA 101A, 2004, 'Guide on Alternative Approaches to Life Safety', Chapter 4, Worksheet 4.7.7).

Worksheet 4.7.7 Individual safety evaluations

Safety parameters	Containment safety (S_1)	Extinguishment safety (S_2)	People movement safety (S_3)	General safety (S_4)
1. Construction	–	–	NA	–
2. Interior finish (corridor and exit)	–	NA	–	–
3. Interior finish (rooms)	–	NA	NA	–
4. Corridor partitions/walls	–	NA	NA	–
5. Doors to corridor	–	NA	–	–
6. Zone dimensions	NA	NA	–	–
7. Vertical openings	–	NA	–	–
8. Hazardous areas	–	–	NA	–
9. Smoke control	NA	NA	–	–
10. Emergency movement routes	NA	NA	–	–
11. Manual fire alarm	NA	–	NA	–
12. Smoke detection and alarm	NA	–	–	–
13. Automatic sprinklers	–	–	/2 =	–
Total Value	$S_1 =$	$S_2 =$	$S_3 =$	$S_4 =$

fire hazards allows a systematic check of potential fire hazards that are in place. The listing of fire protection measures alongside with the potential fire hazards allows a quick check of any safety deficiencies and any need to provide additional fire protection measures to minimize the risk. The checklist method, therefore, is an enumeration of potential fire hazards, fire protection measures, either in place or to be added, and the assessment of the *residual fire risks*. It is used to identify any deficiencies and any corrective measures needed to minimize the fire risks. It does not include, however, the consideration of the logical development of fire events, which will be discussed in Section 5.4 using an event tree.

An example of a checklist method employing quantitative fire risk assessment is shown in Table 5.2. This is the same example that was used in Chapter 4, except that quantitative assessment is employed here rather than qualitative assessment. This example looks at a potential fire hazard in the living room of a house and the consideration of a number

Table 5.2   Quantitative fire risk assessment based on a checklist of potential fire hazards and fire protection measures

No	Fire hazard	Existing fire protection	Inherent fire risk			Additional fire protection	Residual fire risk		
			P (Number of fires per year)	C (Number of deaths per fire)	Risk (Number of deaths per year)		P (multiplication factor of inherent value)	C (multiplication factor of inherent value)	Risk (multiplication factor of inherent value)
1	Fire in living room	Smoke alarms	1.49×10^{-4}	43.2×10^{-3}	6.43×10^{-6}	No smoking material in living room	0.88	1.00	0.88
2	Fire in living room	Smoke alarms	1.49×10^{-4}	43.2×10^{-3}	6.43×10^{-6}	Sprinklers	1.00	0.49	0.49
3	Fire in living room	Smoke alarms	1.49×10^{-4}	43.2×10^{-3}	6.43×10^{-6}	Evacuation drills	1.00	0.40	0.40
4	Fire in living room	Smoke alarms	1.49×10^{-4}	43.2×10^{-3}	6.43×10^{-6}	No smoking material in living room + sprinklers	0.88	0.49	0.43
5	Fire in living room	Smoke alarms	1.49×10^{-4}	43.2×10^{-3}	6.43×10^{-6}	No smoking material in living room + evacuation drills	0.88	0.40	0.35
6	Fire in living room	Smoke alarms	1.49×10^{-4}	43.2×10^{-3}	6.43×10^{-6}	Sprinklers + evacuation drills	1.00	0.20	0.20

Note: In this table, P is the probability of the fire hazard and C is the consequence of the fire hazard. Only the control measure of no smoking material in living room affects the probability of occurrence of fires in the living room. Other fire control measures affect the consequence.

of additional fire protection measures to minimize the risk. Obviously, there could be many potential fire hazards in a house. A complete fire risk assessment would involve the identification of all potential fire hazards and the consideration of various fire protection measures to minimize the risk.

A typical house usually has some fire protection measures, such as smoke alarms. Additional fire protection measures would lower the risk further. Similar to the example in Chapter 4, this example considers six different combinations of three additional fire protection measures. The three additional fire protection measures are: (1) no smoking material (such as cigarettes) in the living room, (2) sprinklers and (3) regular evacuation drills. Each of the three fire protection measures has an impact on either the probability of fire occurrence or the consequence of a fire occurrence. For example, the measure of 'no smoking material in the living room' would have an impact on lowering the probability of fire occurrence; whereas the measures of 'sprinklers' and 'regular evacuation drills' would have an impact on lowering the consequence of a fire occurrence by suppressing or controlling the fire or by allowing the occupants to evacuate more quickly.

It should be emphasized that this is just an example to show how quantitative fire risk assessment can be carried out using a checklist method. There are no standard checklist methods in fire risk assessment.

In Table 5.2, the *inherent fire risk* values (without the help of any fire protection measures) were obtained previously in Chapter 3. Table 3.2 in Chapter 3 shows that the probability of fire occurrence in Canadian houses was 1.75×10^{-3} fires/house/year in 1996 and the percentage of these house fires that occurred in the main living area was 8.5 %. Using these figures, the probability of fire occurrence in the main living area in Canadian houses in 1996 was, therefore, $1.75 \times 10^{-3} \times 8.5$ % or 1.49×10^{-4} fires/house/year. Table 3.2 in Chapter 3 also shows that the consequence of fires originating in the main living area in 1996 was 43.2×10^{-3} deaths/fire, and the resultant risk to life from these fires was 6.43×10^{-6} deaths/house/year. These previously obtained inherent risk values are used in the present example and are shown in Table 5.2.

The *inherent risk* values in Table 5.2 were based on fire statistics which included some fire protection measures, such as smoke alarms, that were required by regulations. If additional fire protection measures are put in place, the inherent fire risks would be further reduced. In Table 5.2, the impact of each of the six fire protection combinations is assessed using a *residual multiplication factor* of the inherent values of the probability or the consequence. This allows the fire protection engineers and the regulators to assess the impact of these fire protection measures based on their assessments of the reduction of the inherent values.

One way to assess the impact of fire protection measures is through the use of statistical information, if they are available. Unfortunately, such information is not always available. The information may be in the databases of collection agencies, but not necessarily in their published reports which usually show general information and not the specific information that is required for fire risk assessment. If no such information is available, subjective judgment may be required. Otherwise, the use of fundamental and rational approach to quantification is required, including the use of mathematical modelling of fire development and occupant evacuation, which will be discussed in later chapters.

For example, there is some statistical information on the benefits of restricting smoking material and of installing sprinklers, but not much information on the benefits of implementing regular evacuation drills. NFPA statistics show that approximately 7 % of fires in homes are caused by smoking materials (NFPA Fire Statistics, 2006) and approximately 14 % of these fires occur in the main living area (Hall, 2006). Therefore, 7 × 14 % or approximately 1.0 % of fires in homes are fires that both originate in the main living area and are caused by smoking material. If these fire statistics can also apply to Canadian homes, then restricting smoking material in the main living area would reduce the number of fire occurrence in the main living area from 8.5 % (see Table 3.2 in Chapter 3) to 7.5 % of house fires. The reduction of fire occurrence from 8.5 to 7.5 % is 12 %. The corresponding residual multiplication factor of the inherent probability value by restricting smoking material is therefore 0.88, which is shown in Table 5.2.

NFPA statistics also show that, based on 1989–1998 data, the reduction in deaths in one and two family dwellings with sprinklers is 51 % when compared with similar dwellings without sprinklers (Kimberly and Hall, 2005). The corresponding residual multiplication factor of the inherent consequence value by installing sprinklers is therefore 0.49, which is shown in Table 5.2.

With regard to the benefits of implementing regular evacuation drills, there is no information on the reduction of death rates that is easily available. For this example, we have to make an assumption. We know that if regular evacuation drills are carried out, there will be faster evacuations and therefore lower death rates. For this example, let us assume a reduction of the death rate by 60 %. The residual multiplication factor of the inherent consequence value by implementing regular evacuation drills is therefore 0.40, which is shown in Table 5.2. In actual fire risk assessments, this value needs to be judged and agreed upon between the fire protection engineers and regulators.

Table 5.2 also shows that the impacts on the consequence of installing sprinklers and of implementing regular evacuation drills are multiplied together. That is, the benefits of sprinklers and of regular evacuation drills have a combined residual consequence factor of 0.49 × 0.40 or 0.20. The multiplication of the residual factors is based on the argument that each fire protection measure reduces the residual death rate by a certain percentage in succession. The death rate is first reduced by the sprinklers suppressing the severity of the fires, and then further reduced by faster evacuation of the occupants.

The reduction of the risk values of the six combinations of additional fire protection measures is shown in Table 5.2. The *residual risk* multiplication factors range from 0.88 to 0.20. The quantification of the risk values allows numerical comparisons of the various fire protection options. This is not the case in qualitative fire risk assessment (see Table 4.4 in Chapter 4).

It should be emphasized again that the values in Table 5.2 are selected by the author as an example to show how such a checklist method can be carried out. These values were selected from available statistical information without much in-depth search. As more statistical information becomes available, more extensive search and detailed analysis are needed to find the correct values. In actual fire risk assessments, these values need to be carefully selected and agreed upon by stakeholders. Subjective judgment of the probabilities and consequences provides a quick assessment of the potential fire risks. More fundamental and rational approaches to quantification, including the use of mathematical modelling of fire development and occupant evacuation, will be discussed in later chapters.

5.4 Event-Tree Method

An *event tree* is another way to identify potential fire hazards, assess their probabilities and consequences, and arrive at risk values. Different from the checklist method, an event tree shows more than a list of potential fire hazards and fire protection measures for the assessment

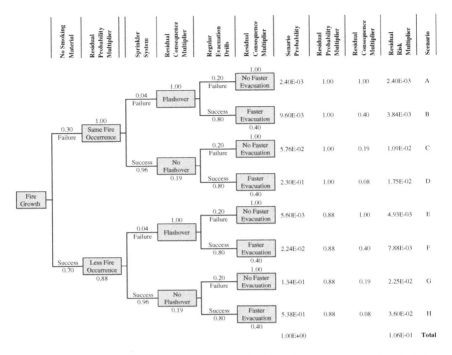

Figure 5.1 An example of an event-tree method for the assessment of probability, consequence and residual risk values for the various fire scenarios in an apartment building.

of the probabilities, consequences and eventually the risk values. The event-tree method involves the construction of an event tree of various *fire scenarios* subsequent to the initiation of a fire hazard, as described in Chapter 2. The fire scenarios provide more logical information for the judgment of probability, consequence and risk values. An example of an event-tree method employing quantitative fire risk assessment is shown in Figure 5.1. This is the same example that was used in Chapter 4, except that quantitative assessment is employed rather than qualitative assessment.

In Figure 5.1, the branching to different events depends on the success or failure of the fire protection measures in place. This example looks at one fire hazard in an assumed apartment building and the consideration of a number of additional fire protection measures to minimize the risk. The same event tree can be constructed for more hazards and more fire protection measures. A complete fire risk assessment would involve the identification of all potential fire hazards and the consideration of various fire protection measures to minimize the risk.

A typical apartment building usually has some fire protection measures, such as fire resistant construction and fire alarms. Additional fire protection measures would lower the risk further. This example considers the same six different combinations of three additional fire protection measures which were considered in the checklist method in Section 5.3. The three additional fire protection measures are: (1) no smoking material (such as cigarettes) in the apartments, (2) sprinklers and (3) regular evacuation drills. Each of the three fire protection measures has an impact on either the probability of fire occurrence or the consequence of a fire occurrence. For example, the measure of 'no smoking material in the apartment' would have an impact on lowering the probability of fire occurrence; whereas the measures of 'sprinklers' and 'regular evacuation drills' would have an impact on lowering the consequence of a fire occurrence by suppressing or controlling the fire or by allowing the occupants to evacuate more quickly.

As was in the discussion of the checklist method, the event tree in Figure 5.1 is only an example to show how an event tree can be used for quantitative fire risk assessment. The description of each event is the judgment by the author for this example only. Other applications may have different descriptions. The descriptions allow more transparent discussions and agreements among stakeholders.

In an event-tree method, the probability of each fire scenario is calculated using the probability values of success or failure of implementing the fire protection measures that are associated with the scenario. Some of these probability values can be obtained from statistics, if they are available. For example, NFPA statistics show that, based on the 1999–2002 data, sprinklers in apartment buildings have a reliability of 96 % of activating and controlling large fires that should activate sprinklers (Kimberly and Hall, 2005). Note that smouldering fires and small fires may not activate sprinklers. If no such information is available, then subjective judgment may be required. For example, there is no statistical information that can be easily found on the probability of success or failure of implementing a 'no smoking material' plan so that there will be a lower rate of fire occurrence. Without such statistical information, we have to make an assumption. Let us assume for this example that the probability of success of implementing a 'no smoking material' plan is 70 %. Similarly, there is no statistical information that can be easily found on the probability of success or failure of implementing a 'regular evacuation drills' plan in apartment buildings so that the occupants would know what to do in case of a fire alarm and would therefore evacuate more quickly than without such drills. Let us assume for this example that the probability of success of implementing a 'regular evacuation drills' plan is 80 %. In real risk assessments, these values need to be carefully analysed and agreed upon by fire safety engineers and regulators. Success is defined as that the fire protection plan actually works.

The scenario probabilities are shown in Figure 5.1. For example, Scenario A has a probability of 2.40×10^{-03}, which is the product of 0.30 (failure probability of implementing a 'no smoking material' plan) $\times$ 0.04 (failure probability of 'sprinkler system') $\times$ 0.20 (failure probability of implementing a 'regular evacuation drills' plan).

In the event-tree method, the probability of fire occurrence for each fire scenario is assessed based on the inherent rate of fire occurrence and the impact of various fire prevention measures to minimize this inherent rate of fire occurrence. In Figure 5.1, the impact of each of the fire protection measures on the inherent rate of fire occurrence is assessed using a *residual probability multiplier*. This allows the fire protection engineers and regulators to assess the impact of these fire protection measures based on their assessments of the reduction of the probability of fire occurrence. Some of these residual probability multipliers can be obtained from statistics, if they are available. If no such information is available, then subjective judgment may be required. For example, there is no statistical information that can be easily found on the reduction of fire occurrence of implementing a 'no smoking material' plan for apartment buildings. Without such statistical information, we have to make an assumption. Let us assume for this example that the residual probability multiplier of a 'no smoking material' plan in apartment building is 0.88, the same as that for house fires (see Table 5.2). That is, the consequence of a successful 'no smoking material' plan is the reduction of the rate of fire occurrence to 0.88 of its inherent value.

This residual probability multiplier for each fire scenario is shown in Figure 5.1. For example, Scenarios A, B, C, D, all with a failure of implementing the 'no smoking material' plan, have a residual probability multiplier of 1 (no reduction); whereas Scenario E, F, G, H, all with a success of implementing the 'no smoking material' plan, have a residual probability multiplier of 0.88. That is, the consequence of a 'no smoking material' plan is the reduction of the rate of fire occurrence to 0.88 of its inherent value.

Also in an event-tree method, the consequence of each fire scenario is assessed based on the inherent consequence of the fire and the impact of the various fire protection measures to minimize the consequence. In Figure 5.1, the impact of each of the fire protection measures on the consequence is assessed using a *residual consequence multiplier*. This allows the fire protection engineers and regulators to assess the impact of these fire protection measures based on their assessments of the reduction of the consequence. Some of these residual consequence multipliers can be obtained from statistics, if they are available. For example, NFPA statistics show that, based on the 1989–1998 data, the reduction in deaths in apartment buildings with sprinklers is 81 % when compared with similar buildings without sprinklers (Kimberly and Hall, 2005). The residual consequence multiplier of a sprinkler system therefore is 0.19. That is, the consequence of installing a sprinkler system is the reduction of the death rate per fire to 0.19 of its inherent value. If no such information is available, then subjective judgment may be required. For example, there is no statistical information that can be easily found on the death reduction benefit of implementing a 'regular evacuation drills' plan. Without such statistical information, we have to make an assumption again. Let us assume for this example that the residual consequence multiplier of a 'regular evacuation drills' plan is 0.40. That is, the consequence of a 'regular evacuation drills' plan is the reduction of the death rate per fire to 0.40 of its inherent value.

This residual consequence multiplier for each scenario is shown in Figure 5.1. For example, Scenario B has a residual consequence multiplier of 0.40, which is the product of 1.00 (residual consequence multiplier of a failed sprinkler system) × 0.40 (residual consequence multiplier of a successful 'regular evacuation drills' plan).

Figure 5.1 shows the residual risk values of all the fire scenarios which are based on the success or failure of three fire protection measures. The probability value of each fire scenario is the product of the individual probability values of all the branches that are associated with that scenario. The residual probability multiplier of each scenario is the product of the individual probability multipliers of all the fire protection measures that are associated with that scenario. The residual consequence multiplier of each scenario is the product of the individual consequence multipliers of all the fire protection measures that are associated with that scenario. Finally, the residual risk multiplier for each scenario is the product of (scenario probability) × (residual probability multiplier) × (residual consequence multiplier). For example, Scenario E has a scenario residual risk multiplier of 4.93×10^{-03}, which is the product of 5.60×10^{-03} (scenario probability) × 0.88 (residual probability multiplier) × 1.00 (residual consequence multiplier).

In Figure 5.1, the multiplication of the residual multipliers is based on the argument, as discussed in Section 5.3, that each fire protection measure reduces the rate of fire occurrence, or the severity of the fire, or the death rate per fire, in succession by a certain percentage.

It should be emphasized again that the values in the example are selected by the author as an example to show how such event-tree method can be carried out. In real risk assessments, these values need to be carefully analysed and agreed upon by fire safety engineers and regulators. Subjective judgment of the probabilities and consequences provides a quick assessment of the potential fire risks. More fundamental and rational approaches to quantification, including the use of mathematical modelling of fire development and occupant evacuation, will be discussed in later chapters.

Different from the checklist method, the event-tree method allows the summation of the risk values of all the fire scenarios into one single risk value for the whole system. This allows direct comparisons of the risk values of various fire safety design options, including code-compliant designs. Figure 5.1 shows the combined residual risk multiplier of implementing these three fire protection measures is 1.06×10^{-1}. That is, the residual risk is reduced to 10.6 % of its inherent value.

The inherent fire risk values of apartment buildings can be obtained from statistics. For example, in Canada, the 1996 Canadian fire statistics (Council of Canadian Fire Marshals and Fire Commissioners, 1996) show that the total number of fire deaths in apartment buildings in that year was 88. Also, the 1996 Canadian census data (Statistics Canada, 1996) show that the total number of apartment units under the category of 'apartment 5 or more storeys' was 979 470. The risk of dying in an apartment unit, therefore, was 88 deaths divided by 979 470 apartments or 8.98×10^{-5} deaths/apartment/year. Compare this risk value with that of house fires of 1.94×10^{-5} deaths/house/year (see Chapter 3, Section 3.3.3), the risk of apartment fires is much higher. Part of the explanations could be that the fire risks in a house are caused by fires originating in the same house, whereas the fire risks in an apartment unit are caused by fires originating from all apartment units in a building.

5.5 Summary

Quantitative fire risk assessment is an assessment involving numerical quantifications of both the probability of occurrence of a fire hazard, or fire scenario, and the consequence of that fire hazard or fire scenario. The multiplication of the numerical values of probability and consequence gives each fire scenario a numerical fire risk value. The cumulative sum of the risk values from all probable fire scenarios gives an overall fire risk value. The assessed risk can be risk to life, loss of property and so on. Quantitative fire risk assessment allows a numerical comparison of the overall fire risk values of different fire safety designs in a building. It also allows the assessment of equivalency by comparing the fire risk of an alternative fire safety design with that of a code-compliant design.

There are in general two ways of conducting systematic quantitative fire risk assessments: (1) using a checklist to go through a list of potential fire hazards and the quantitative assessment of their fire risks; (2) using an event tree to go through a set of the potential fire scenarios and the quantitative assessment of their fire risks. Within the checklist method, there are specific methods that have been developed by various organizations for their own use. One particular one is called the *risk indexing* method which uses well-defined schedules, or tables, to rate the risks. In both the checklist and event-tree methods, the outcome is a list of potential fire hazards, or fire scenarios, and their assessed fire risk values. Summation of all these individual risk values gives an overall fire risk value in a building that can be used for comparisons with those of alternative fire safety designs.

It should be noted that there are semi-quantitative assessments, where only one of the two parameters (probability or consequence) is assessed quantitatively. The other parameter that is not assessed quantitatively is assessed qualitatively (see Chapter 4). This type of assessment is neither qualitative nor quantitative. In this chapter, we discussed only quantification of both the probability and consequence parameters. The quantification of both the parameters was based on statistical data if they are available, or subjective judgment if such data are not available. More fundamental and rational approaches to quantification, including the use of mathematical modelling, will be discussed in later chapters.

5.6 Review Questions

5.6.1 Use the checklist method in Table 5.2 to do a quantitative fire risk assessment for a night club fire. Change 'no smoking material' to 'no firework on stage' and change 'evacuation drills' to 'prior evacuation instructions before show starts'. Make your assumption on what the residual multiplication factors should be. Review the night club fire discussion in Chapter 3.

5.6.2 Use the event-tree method in Figure 5.1 to find the overall residual risk multiplier for a fire protection system that includes 'sprinklers'

protection only, but not 'no smoking material' or 'regular evacuation drills' fire protection measures. Change both the failure
probabilities of 'no smoking material' and 'regular evacuation
drills' to 100 %.

5.6.3 Use the event-tree method in Figure 5.1 to find the residual
risk multiplier for a fire protection system that does not include
'sprinkler' protection but has 'no smoking material' and 'regular
evacuation drills' fire protection measures. Compare the overall
residual risk multiplier with that in 5.6.2. This is a comparison
of the risk reduction of either installing a sprinkler system, or
implementing better fire prevention or occupant evacuation plans.
Change the failure probability of sprinklers to 100 %.

References

Council of Canadian Fire Marshals and Fire Commissioners (1996) Fire
    Losses in Canada, Annual Report, Table 3b, http://www.ccfmfc.ca/stats/
    en/report_e_96.pdf

Hall, J.R., Jr. (2006) *The Smoking-Material Fire Problem*, NFPA Report, August
    2006, Table 6: Smoking-Material Fires in Homes, by Area of Fire Origin,
    NFPA, Quincy, MA.

Kimberly, D.R. and Hall, J.R., Jr. (2005a) *U.S. Experience with Sprinklers and
    other Fire Extinguishing Equipment*, NFPA Report, August 2005, Table 9:
    Reduction in Civilian Deaths, NFPA, Quincy, MA.

Kimberly, D.R. and Hall, J.R., Jr. (2005b) *U.S. Experience with Sprinklers and
    other Fire Extinguishing Equipment*, NFPA Report, August 2005, Table 7B:
    Sprinklers Performance, NFPA, Quincy, MA.

NFPA 101A (2004) *Guide on Alternative Approaches to Life Safety*, NFPA,
    Quincy, MA.

NFPA Fire Statistics (2006) *Leading Causes of Structured Fires in Homes,
    1999-2002 Annual Averages*, NFPA, Quincy, MA.

Statistics Canada (1996) Census Canadian Private Households by Size, http://
    www.statcan.ca/english/census96/oct14/hou.pdf

Watts, J.M., Jr. (2002) Fire risk indexing, *SFPE Handbook of Fire Protection
    Engineering*, 3rd edn, Section 5-10, NFPA, Quincy, MA.

Part II

Fundamental Approach to Fire Risk Assessment

6

Fundamental Approach to Fire Risk Assessment

In previous chapters, simple fire risk assessment methods were discussed. These methods include those where the assessment is made by comparing the present fire situation with past fire experience or incident data; and those where the assessment is made by utilizing simple checklist or risk matrix methods. The problem with the use of past experience or incident data is that the past and the present situations may not be exactly the same and therefore the past experience or incident data may not be applicable to the present situation. The problem with the use of checklist and risk matrix methods is that they are based on subjective judgments which may or may not be correct and can not be verified. One person's judgment may be different from those by others. The same person may not make the same judgment consistently in similar situations.

A better way to conduct fire risk assessment is to use the fundamental approach. Fundamental approach involves: (1) the construction of all possible fire scenarios that a fire initiation may develop into; (2) the construction for each fire scenario of a sequence of fire events that follow the course of an actual fire development; and (3) the mathematical modelling of these fire events to predict the outcome of occupant fatalities and property loss. The various fire scenarios that a fire initiation can develop into are governed, as was discussed in previous chapters, by the success and failure of fire protection measures. The sequence of fire events that follows the course of an actual fire development includes fire growth, smoke spread, occupant evacuation and fire department response. The fundamental approach is to follow the logical development of these

Principles of Fire Risk Assessment in Buildings   D. Yung
© 2008 John Wiley & Sons, Ltd

fire events. Details of this approach will be discussed in the following chapters of this book.

The construction of a complete set of possible fire scenarios is showed in Figure 6.1. The construction is based on connecting the possible fire scenarios that can be constructed in each of the five basic fire barriers. The five basic fire barriers were discussed previously in Chapter 2. The possible fire scenarios in each of the five basic fire barriers are based on the success and failure of the fire protection measures that are employed in each of these barriers, as shown in Figure 6.2.

For each fire scenario, the sequence of events that can lead to occupant deaths is shown in Figure 6.3, whereas the sequence of events that can lead to property loss is shown in Figure 6.4. The fire events include the fast fire events of fire growth, smoke spread, occupant evacuation and fire department response, and the slower fire event of fire spread which spreads by breaching the fire resistant boundaries one at a time. The fast events are used to determine whether occupants are trapped in a building, depending on whether they can evacuate in time before the arrival of the critical smoke conditions in the egress routes that prevent evacuation; and whether the trapped occupants can be rescued by the firefighters, depending on the fire department's response time and resources. Both smoke spread and the slower fire spread cause deaths to

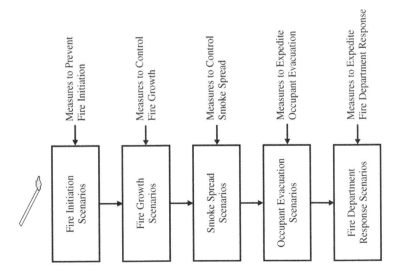

Figure 6.1 A complete set of possible fire scenarios can be constructed based on connecting the possible fire scenarios that can be constructed in each of the five basic fire barriers.

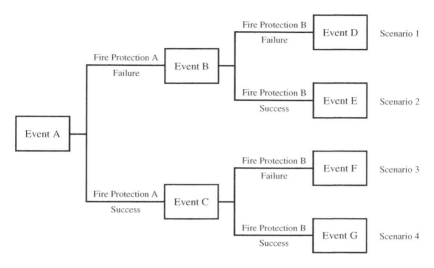

Figure 6.2 Fire scenarios in each of the five basic fire barriers can be constructed based on the success and failure of fire protection measures that are employed in these basic fire barriers.

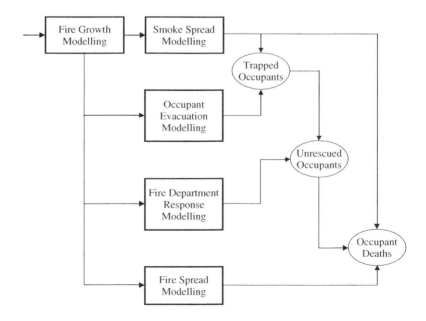

Figure 6.3 Fire events that can lead to occupant deaths.

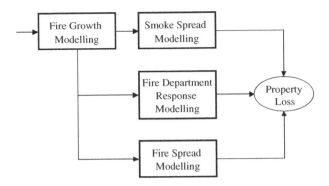

Figure 6.4   Fire events that can lead to property loss.

those occupants who are trapped and can not be rescued by the fire-fighters. Both smoke spread and the slower fire spread cause loss to the property. Details of this approach will be discussed in the following chapters of this book.

Note that the fire initiation event is not included in Figure 6.3. The fire initiation event is a random event. The normal fire protection measure to prevent this fire event from happening is fire safety education, such as educating people about the danger of cigarettes and matches as an ignition source.

In the fundamental approach to fire risk assessment, the probabilities of fire scenarios are governed by the reliability parameters of the fire protection measures and the outcome of the models are governed by the input parameters. The values for some of these parameters are known, but for some others, their values are not known. For these parameters whose values we don't know, subjective judgments of their values have to be made. With this fundamental approach, however, the judgment is made on parameters that are very basic and can be more likely agreed upon by fire protection engineers and regulators. For example, instead of making a judgment on whether sprinkler protection can improve safety by a certain percentage, the judgment is only made on the reliability of the sprinkler system that will be activated in the event of a fire. Or, instead of making a judgment on whether adding another emergency exit will improve safety by a certain percentage, the judgment is only made on the percentage of occupants who will respond to a fire alarm. All of these will be discussed in detail in the following chapters of this book.

The added benefit of a fundamental approach is that it allows us to identify those basic parameters whose values we don't know well. The identification of those parameters often leads to research activities to determine their values. As more research leads to better understanding of these unknown parameters, less judgment of their values is needed. The assessment becomes more and more accurate.

7

Fire Growth Scenarios

7.1 Overview

Several input parameters govern how a fire develops in an enclosed space in a large building which in this context is termed a *compartment*. These parameters can be both deterministic and random. Deterministic parameters include:

- fuel type
- fuel load
- compartment geometry and
- ventilation conditions.

These parameters can be determined before a fire safety design or fire risk assessment is carried out. Random parameters are those that can not be determined a priori and include:

- ignition source
- ignition location
- fuel arrangement.

As a consequence of these random parameters, many types of fires can develop, from small fires to flashover fires. Rather than considering these parameters individually, which would result in a large number of possible fire growth scenarios, an alternative approach is to consider the

Principles of Fire Risk Assessment in Buildings   D. Yung
© 2008 John Wiley & Sons, Ltd

types of fires that have occurred in the past as a result of these random parameters:

1. smouldering fires,
2. non-flashover flaming fires
3. flashover fires.

We can set up fire growth scenarios based on these three fire types and can include automatic suppressions in the fire growth scenarios based on the probabilities that suppressing the fires may succeed or fail. By using fire growth scenarios along with the output from fire models under different fire growth scenarios we can obtain the necessary information to assess fire risks to a building's occupants and properties.

7.2 Compartment Fire Characteristics

In a large building, the built space is typically divided into a number of enclosed spaces, or compartments. For example, in an apartment building, the built space is divided into a number of apartment units, with each unit further divided into a number of rooms. Fires in a building, therefore, usually begin in an enclosed space, or compartment. The compartment can be a small one, such as a room in an apartment unit, or a large one, such as an atrium in an office building.

The development of a fire in a compartment is not only governed by the flammability characteristics of the burning object, but is also influenced by the environmental conditions in the compartment within which the fire is developing. The development of a fire, in turn, also changes the environmental conditions in the compartment. The fire and the compartment are, therefore, intertwined with each other. This is especially true in small compartments where a fire development can quickly change the environmental conditions in the compartment. In larger compartments, the time required to change the environmental conditions in the compartment is longer. How much longer depends on the size of the fire and that of the compartment. Consequently, in larger compartments, fire development is not affected by the environmental conditions until some time later when the conditions are changed by the fire.

Figure 7.1 is an illustration of a *compartment fire* (from Yung and Benichou, 1999). In this figure, the fire begins with the ignition of an upholstered chair. The burning of the chair forms a fire plume which spreads heat, smoke and toxic gases upward into the ceiling. Eventually, a hot smoke layer is formed at the ceiling. As the hot layer builds up

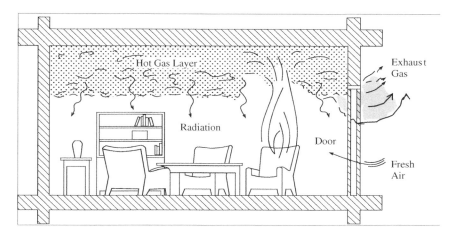

Figure 7.1 Illustration of a fire development in a compartment (from Yung and Benichou, 1999, reproduced by permission of the Fire Protection Research Foundation).

and becomes thicker, it reaches down to the top of the door opening, spilling out hot and toxic gases from the top of the door opening. The exhaust gases can spread very quickly to other parts of the building, posing risks to the occupants and properties in the building. To make up for the exhaust gases, fresh air enters the compartment through the bottom of the door. The fresh air provides a fresh supply of oxygen that is needed to sustain the burning. In the meantime, the hot layer also emits radiant heat back to the combustibles below, intensifying the fire and enhancing the spread of the fire to other combustibles. As the fire develops, the compartment gets hotter and hotter. If the temperature in the hot layer reaches a critical temperature of about 600 °C, the radiant heat from the hot layer is such that it can cause most combustibles in the compartment to ignite spontaneously, forming a flashover fire. If the fire does not produce enough heat to reach the flashover temperature, usually as a result of not being able to spread to other combustibles, the fire would not get to flashover and would decay soon after reaching a certain level of fire intensity.

The severity of a fire is usually measured by how much heat and toxic gases it releases when it burns. Figure 7.2 shows the typical *heat release rate* profiles of flashover and non-flashover fires. *Flashover fires* are those that reach the flashover conditions, continue to burn as fully developed fires and then decay when most of the combustibles have been consumed. *Non-flashover fires* are those that do not reach the flashover conditions. They usually reach some intermediate levels of heat release

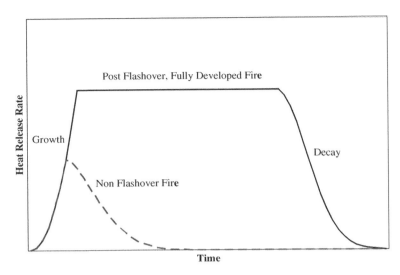

Figure 7.2   Typical heat release rate curves of flashover and non-flashover fires.

rate and then die down. A third type of fire that is not shown in the figure is the *smouldering fire*. Smouldering fires are those that burn slowly, producing mainly smoke and toxic gases but not much heat.

Some examples of fire experiments that show the compartment effect on fire development are shown in Figures 7.3 and 7.4 (from Yung, Wade and Fleischmann, 2004). These experiments were originally conducted at the University of Canterbury in New Zealand (Denize, 2000; Girgis, 2000). Figure 7.3 shows the heat release rates (HRR) from the burning of two identical chairs, with one in the open under a furniture calorimeter (Babrauskas, 2002) and the other in a standard ISO 9705 compartment (ISO, 1993). The chair that was burned in a compartment generated a much higher heat release rate as a result of the heat feedback from the compartment. Similarly, Figure 7.4 shows the production of CO and CO_2 from the burning of the same two identical chairs, again with one in the open under a furniture calorimeter and the other in a standard ISO 9705 compartment. The chair that was burned in a compartment produced much higher CO and CO_2 concentrations, especially CO, as a result of oxygen vitiation in the compartment which changed the chemistry of the combustion.

Fire development in a compartment is very complex because of the interaction of the fire and the environmental conditions in the compartment. Fire research in the past 20 years, however, has produced a number of advanced computer fire models that can model this complex

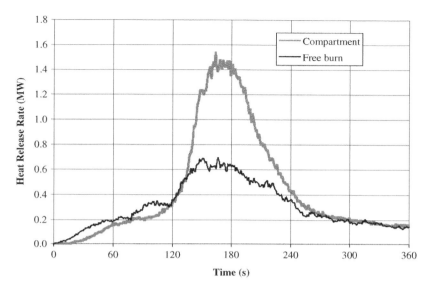

Figure 7.3 Heat release rates from two identical chairs: one burned in the open under a furniture calorimeter and the other in a standard ISO 9705 compartment (from Yung, Wade and Fleischmann, 2004, reproduced by permission of the Society of Fire Safety, Australia).

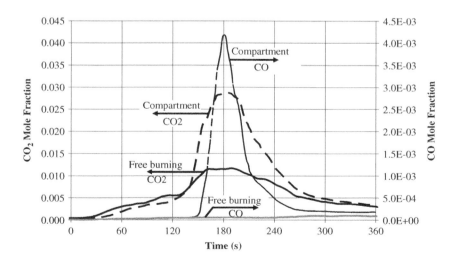

Figure 7.4 CO_2 and CO concentrations (measured in the exhaust duct) from two identical chairs: one burned in the open under a furniture calorimeter and the other in a standard ISO 9705 compartment (from Yung, Wade and Fleischmann, 2004, reproduced by permission of the Society of Fire Safety, Australia).

fire growth in a compartment. These computer models include FDS (McGrattan and Forney, 2006), CFAST (Peacock *et al.*, 2005) and BRANZFIRE (Wade, 2002). All of these models require user specification of the values of the input parameters to run. Input parameters to these models govern the course of the fire development. Output parameters from these models describe the outcome of the fire development.

In this chapter, we will discuss how to manage the input parameters as well as how to use the output parameters for fire risk assessment. Input parameters to these models can be deterministic or random parameters. Output parameters from these models provide the information that can be used for fire risk assessments in both the compartment of fire origin as well as in other locations in a building. We will not, however, discuss fire modelling itself which has already been the subject of many excellent books and publications (Drysdale, 1998; Karlsson and Quintiere, 2000; SFPE, 2002).

7.3 Fire Model Input and Output Parameters

In this section, we will discuss how to manage fire model input parameters and how to use fire model output parameters for fire risk assessment. Input parameters to fire models govern the course of the fire development; whereas output parameters from fire models describe the outcome of the fire development.

Input parameters can be both deterministic and random. *Deterministic parameters* are those that can be determined before a fire safety design or fire risk assessment is carried out. For example, the geometry of the compartment of fire origin is known before a fire safety design or fire risk assessment is carried out. *Random parameters* are those that can not be determined a priori. For example, the location of the ignition point is not known before a fire safety design or fire risk assessment is carried out. Output parameters from these models provide the information that can be used for fire risk assessments in both the compartment of fire origin as well as in other locations in a building. All of these parameters are discussed in more detail in the following sections.

7.3.1 Deterministic Input Parameters

Deterministic input parameters are those that can be determined before a fire safety design or fire risk assessment is carried out. They include fuel type, fuel load, compartment geometry and ventilation conditions (Yung, Wade and Fleischmann, 2004; Yung and Benichou, 2002).

7.3.1.1 Fuel Type

Fuel type is the type of combustibles in the compartment, such as the upholstered furniture in an apartment unit or the office furniture in an office room. The *flammability properties* of these combustibles affect the type of fire that will develop, whether smouldering or flaming and whether slow or fast. They also affect the amount of heat and the quantity of smoke and toxic gases that are generated. Table 7.1 shows the important flammability properties of some of the common materials for well-ventilated fires. For other materials, consult the reference (Tewarson, 2002, Table 3–4.14) which contains the flammability properties of an extensive list of materials. For under-ventilated fires, the values of these properties can be much higher, as a result of incomplete combustions. In this table, ΔH_c is the *maximum heat of combustion* which gives the theoretical maximum amount of heat that can be generated per unit fuel burned. The parameter ΔH_{eff} is the *effective heat of combustion* which gives the effective amount of heat that is generated per unit fuel burned. The ratio of $\Delta H_{eff}/\Delta H_c$ is the *combustion efficiency*. The parameter ΔH_{rad} is the *flame radiant loss* which gives a measure of the amount of heat that is radiated from the flame inside the compartment. The remaining heat stays with the exhaust flow out of the compartment. The ratio of $\Delta H_{rad}/\Delta H_{eff}$ gives the *radiant loss*

Table 7.1 Flammability properties of some common materials for well-ventilated fires (source: Tewarson, 2002, Table 3–4.14).

Material	ΔH_c $(kJ\,g^{-1})$	ΔH_{eff} $(kJ\,g^{-1})$	ΔH_{rad} $(kJ\,g^{-1})$	y_{CO2} (g/g)	y_{CO} (g/g)	y_s (g/g)
Wool	–	19.5	–	–	–	0.008
Wood (pine)	17.9	12.4	3.7	1.33	0.005	0.015[a]
Propane	46.0	43.7	12.5	2.85	0.005	0.024
Kerosene	44.1	40.3	14.1	2.83	0.012	0.042
Nylon	30.8	27.1	10.8	2.06	0.038	0.075
Polyester	32.5	20.6	9.8	1.65	0.070	0.091
Polyurethane foam (flexible)	26.2	17.8	9.2	1.55	0.010	0.131
PVC	16.4	5.7	2.6	0.46	0.063	0.172

In the table, ΔH_c is the maximum heat of combustion per unit of fuel burned; ΔH_{eff} is the effective heat of combustion per unit of fuel burned; ΔH_{rad} is the flame radiant loss per unit of fuel burned; y_{CO2} is the yield of CO_2 per unit of fuel burned; y_{CO} is the yield of CO per unit of fuel burned; and y_s is the yield of soot (smoke particles) per unit of fuel burned.
[a] For red oak (y_s for pine not available); – = not measured or negligible.

fraction. The yield parameters y_{CO2}, y_{CO} and y_s give the amounts of CO_2, CO and soot (smoke particles) that are generated per unit of fuel burned. The yield parameters affect the concentrations of CO_2, CO and soot in the compartment as well as in the exhaust flow leaving the compartment.

7.3.1.2 Fuel Load

Fuel load is the amount of fuel in the compartment, for example, the amount of furniture in an apartment unit. It affects both the severity of the fire and the duration of the fire. Fuel load is usually characterized by the loading density, expressed in $kg\,m^{-2}$ or $MJ\,m^{-2}$. It is usually randomly distributed. The random nature of the distribution is discussed in the Section 7.3.2.

7.3.1.3 Compartment Geometry and Properties

Compartment geometry and properties refer to the dimensions and shape of the compartment and the thermal properties of the compartment boundaries. They affect both the build-up of heat within the compartment as well as the radiant feedback from the compartment boundaries to the fuel.

7.3.1.4 Ventilation Conditions

Ventilation includes both natural, such as window and door openings, and mechanical, such as fans. Ventilation depends on the conditions of these windows and doors (open or closed) and fans (on or off). *Ventilation conditions* affect the supply of fresh air into the compartment and, consequently, the fire development within the compartment. They also affect the outflow of toxic gases which affect the occupants in the building.

7.3.2 Random Input Parameters

Random input parameters are those that can not be determined before a fire safety design or fire risk assessment is carried out. They include the *ignition source, ignition location* and the *fuel arrangement,* or distribution, of the fuel load. As a result of these random parameters, fires can develop into different types: from a small fire to a flashover fire. Some of the ways to deal with these random parameters will be discussed in the Section 7.4.

7.3.3 Output Parameters for Fire Risk Assessment

Output parameters are those that describe the outcome of the fire development in a compartment. Fire development in a compartment is very complex and the modelling of these fires is usually conducted using fire models (mathematical or computer models). The output from these fire models provides the information that can be used to assess the fire risks in the compartment of fire origin as well as in other locations in a building. Output parameters that are important include the *temperature*, CO, CO_2 and *soot concentrations*, both within the compartment of fire origin as well as in the exhaust gases leaving the compartment. Also important is the *exhaust flow rate* which is a measure of the rate of spread of the heat, toxic gases and smoke into other locations in a building. Together, these parameters provide the information that can be used to assess the fire risks to the occupants and properties in a building.

As mentioned previously, fire research in the past 20 years has produced a few advanced computer fire models that can model this complex fire growth in a compartment. These models include FDS (McGrattan and Forney, 2006), CFAST (Peacock *et al.*, 2005), and BRANZFIRE (Wade, 2002). Examples of the output of the FDS are shown in Figures 7.5 and 7.6. Figure 7.5 shows the temperature profile in a compartment fire, whereas Figure 7.6 shows the velocity profile.

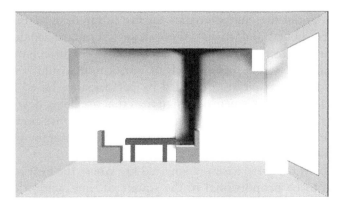

Figure 7.5 FDS (Fire Dynamic Simulator) output of the temperature profile in a compartment fire (figure courtesy and by permission of Dr Yunlong Liu of Sydney, Australia).

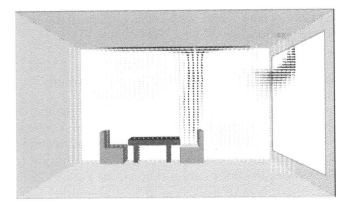

Figure 7.6 FDS (Fire Dynamic Simulator) output of the velocity profile in a compartment fire (figure courtesy and by permission of Dr Yunlong Liu of Sydney, Australia).

7.4 Design Fires

Design fires are prescribed fires that can be used by fire protection engineers for performance-based fire safety designs or fire risk assessments in buildings. Design fires characterize the fire growth in the compartment of fire origin. Different design fires are used for different occupancies because the combustibles are different. For example, design fires for residential buildings are different than those for office buildings.

Design fires are being developed by standards organizations (BSI, 2003; ISO, 2003) and research organizations (Yung, Wade and Fleischmann, 2004; Hadjisophocleous and Zalok, 2004; Bwalya *et al.*, 2006) so that the initial fire characteristics for fire safety designs and fire risk assessments can be standardized. Without such standardized design fires, different fire protection engineers may use different fire characteristics for their fire safety analyses. A lack of uniformity would result in the levels of fire safety that are provided by performance-based fire safety designs or fire risk assessments. Design fires that are being developed can range from a very simple time-dependent heat release rate curve to those that take into considerations of all of the governing input parameters that were discussed in the previous section.

The simplest design fire is the *t*-squared fire where the heat release rate is assumed to be proportional to the square of time, with a coefficient that can be assigned different values, depending on the burning material, to denote different growth rates of the fire development. The *t*-squared fire is expressed as:

$$\dot{Q} = \alpha t^2, \tag{7.1}$$

where $\dot{Q}$ is the heat release rate in kW, α is the fire growth rate coefficient in $kW \cdot s^{-2}$ and t is the time in s. *T-squared fires* use four classifications of fires to represent all fire growth rates: slow, medium, fast and ultra fast (BSI, 2003, Part 1). Table 7.2 shows the four fire growth rates with their corresponding α values. Table 7.2 also shows the characteristic times required for the four fire growth rates to reach a fire intensity of 1 MW. A slow fire would take 600 seconds to reach a 1 MW fire, whereas an ultra-fast fire would take only 75 seconds.

For fire safety analyses, different fire growth rates are recommended for different occupancies. Table 7.3 shows, as an example, the fire growth rates for different occupancies that are recommended by the British Standards (BSI, 2003, Part 1) for fire safety designs and fire risk assessments.

T-squared fires are simple heat release rate curves. They do not produce all the fire characteristics that are needed for fire safety designs or fire risk assessments. These fire characteristics include time-dependent development of temperature, smoke, CO and CO_2 concentrations and mass flow rates in the compartment of fire origin, as well as in the

Table 7.2   *T*-squared fires (BSI, 2003, Part 1).

Growth rate	α ($kW \cdot s^{-2}$)	Time (s) to reach 1 MW
Slow	0.003	600
Medium	0.012	300
Fast	0.047	150
Ultra fast	0.188	75

Table 7.3   *T*-squared fires (source: BSI, 2003, Part 1).

Occupancy	Fire growth rate
Picture gallery	Slow
Dwelling	Medium
Office	Medium
Shop	Fast
Industrial storage	Ultra fast

exhaust gases that leave the compartment of fire origin. These output parameters are usually calculated separately, based on the assumed HRR and the governing input parameters. Heat release rates, however, are affected by the fire conditions in the compartment. The coupling of the heat release rate and the fire conditions in the compartment can only be handled by computer fire models.

A number of research organizations are trying to come up with standardised design fires for fire safety engineers to use. There are basically two ways to develop these fires. One way is to measure the fire growth characteristics using fire experiments with a representative fuel arrangement. The other is to identify suitable input parameters for use with fire models. These two ways of developing design fires are discussed in the following two sections.

7.4.1 Based on Fire Experiments with Representative Fuel Arrangements

One way to develop a design fire for a specific occupancy is to measure the fire growth characteristics in a test compartment using a *representative fuel arrangement*. In this approach, the test conditions and the representative fuel arrangement are assumed to represent all of those input parameters that govern fire development in that occupancy. Input parameters that govern fire development, which include both deterministic and random parameters, were discussed in Section 7.3. In this approach, the representative fuel arrangement is assumed to represent the fuel type, fuel load and fuel arrangement in that occupancy; the test procedure is assumed to represent the ignition source and the ignition point in that occupancy; and the test compartment is assumed to represent the compartment geometry and ventilation conditions in that occupancy. In fire safety designs and fire risk assessments, there are usually many design fire scenarios that need to be considered, as a result of the many combinations of the probable values of the governing input parameters. In this simplified approach, the many fire scenarios are assumed to be represented by only one design fire scenario.

One example of the use of the representative fuel arrangement approach is the one proposed by Hadjisophocleous and Zalok (2004). They used a representative fuel package to experimentally produce a design fire for use for the storage areas of retail stores in multi-storey office buildings. Retail stores, such as office supplies, gift shops, drug stores, electronic and so on, are usually located on the ground, or underground, levels of multi-storey office buildings. The storage areas of these retail stores contain typically large amounts of combustibles, or fire loads, which pose significant fire hazards to the occupants in a multi-storey office building should a fire occur in these areas. A proper design fire for these areas is needed to allow for credible fire safety analyses.

In their approach, Hadjisophocleous and Zalok (2004) first conducted a careful survey and analysis of the fire loads, and the types of combustibles, in such storage areas of retail stores in multi-storey office buildings. They found that the representative fire load density, in calorific values per square metre of floor area, is $2320\,MJ\,m^{-2}$. They chose the 90-percentile value of the probability distribution in the survey as the representative fire load density. They also found that the various combustibles that contribute to the fire load, and their respective percentage contributions, are: wood and paper (49.1 %), plastics (31.1 %), food (5.7 %), textiles (5.6 %) and others (8.5 %).

They then constructed a representative fuel package for the fire tests, based on the above percentages of combustibles. The fuel package they assembled weighed 102 kg, had a base area of 1 m × 1 m and a total calorific value of 2320 MJ. They burned the fuel package in an ISO 9705 standard room to generate the fire growth characteristics for use as a representative design fire for such storage areas. The results showed the heat release rate reached a peak value of 1.39 MW and the temperature in the hot layer in the room reached a peak value of 662 °C, both within 8 minutes. Other important parameters that are needed for fire risk assessments, such as exhaust flow rate, temperature, smoke, CO and CO_2 concentrations, were also recorded in the exhaust duct. For more details of their results, consult the paper by Hadjisophocleous and Zalok (2004).

Another example of the use of the representative fuel arrangement approach is the one proposed by Bwalya et al. (2006). They used a representative furniture arrangement to generate fires for fire performance tests in houses. In their approach, they first conducted a survey of the fire loads in the living areas in Canadian homes. They found that the typical fire load is $350\,MJ\,m^{-2}$. Note that this value is much lower than the value of $780\,MJ\,m^{-2}$ that is recommended by the British Standards (BSI, 2003, Part 1) for fire safety designs and fire risk assessments for dwellings, which is shown in Table 7.4 later in this chapter.

They then proposed a furniture arrangement consisting of a three-seater sofa, two small wood cribs underneath the sofa and two large wood cribs on two sides of the sofa. All of these items were arranged on a base area of $8.3\,m^2$. The three-seater sofa has a steel

Table 7.4 Fire load densities for various occupancies (source: BSI, 2003, Part 1).

Occupancy	Fire load density (MJ m^{-2})	
	Mean	90 Percentile
Hotel bedroom	310	460
Offices	420	670
Dwelling	780	920
Shops	600	1100
Libraries	1500	2550
Hospital storage	2000	3700

frame construction with three polyurethane foam seats and three polyurethane foam backs. The polyurethane foams are all 610 mm long by 610 mm wide. The seats have a thickness of 100 mm whereas the backs have a thickness of 150 mm. Each of the two small wood cribs weighs 25.9 kg and each of the two large wood cribs weighs 50.9 kg. Based on their calorimeter tests, the proposed furniture arrangement has a total fire load density of 350 MJ m^{-2}.

The proposed furniture arrangement is expected to follow a fire growth rate between a fast and an ultra-fast t-squared fire (see Table 7.2) and be able to reach a 2 MW peak heat release rate in less than 200 seconds. For more details of their results, consult the report by Bwalya *et al.* (2006). Although their furniture arrangement is proposed mainly for use in a test facility designed for fire performance tests for houses, such furniture arrangement can also be used in standard room burn facilities to generate the design fires for use for residential buildings, such as apartment buildings.

7.4.2 Based on Suitable Input Parameters for Use with Fire Models

Instead of using a representative fuel arrangement, a more general approach is to identify proper values for the governing input parameters for different occupancies and then input these values into computer fire models to generate the design fires. Input parameters that govern fire development, which include both deterministic and random parameters, were discussed in Section 7.3. This approach of identifying proper values for the governing input parameters was proposed by Yung, Wade and Fleischmann (2004) for apartment buildings. Their treatment of the input parameters, both deterministic and random as described in Section 7.3, is discussed below.

7.4.2.1 Fuel Type

This governing parameter affects the type of fire that will develop, whether smouldering or flaming and whether slow or fast. It also affects the amount of heat and the quantity of smoke and toxic gases that are generated. In any occupancy, such as an apartment unit, there is usually more than one type of fuel. If a fire occurs in such an apartment unit, the fire burns with the characteristics of not just one single type of fuel, but those of many types of fuel. The flammability properties of the fuel, such as those shown in Table 7.1, affect the burning behaviour. Yung, Wade and Fleischmann (2004) proposed to survey all available compartment fire databases to identify proper flammability properties for use for apartment buildings. In essence, they proposed to identify proper flammability values, such as those in Table 7.1, that represent the mix of combustibles in an apartment unit.

Yung, Wade and Fleischmann (2004) also proposed that, for flaming fires in apartment buildings, the rate of fire growth should be identified through the survey of all available compartment fire databases. The survey would show the appropriate fire growth rate for apartment buildings which could very well be different than the medium fire that is recommended by the British Standards, as shown in Table 7.3. Already, the proposed design fire by Bwalya et al. (2006) for residential buildings, as was discussed in Section 7.4.1, follows a fire growth rate that is between fast and ultra-fast t-squared fires. Other furniture calorimeter experiments have also shown furniture fires to be between fast and ultra-fast fires. Figure 7.7 (reproduced here from Yung, Wade and Fleischmann, 2004) shows the HRR of five chairs of identical design but with different fabric coverings. These experiments were originally conducted at the University of Canterbury in New Zealand (Enright and Fleischmann, 1999). The initial HRR follow fire growth rates that are between fast and ultra-fast t-squared fires.

Fires in apartment units can also occur as smouldering fires. Smouldering fires are slow burning fires. The British Standards (BSI, 2003, Part 1) recommend the following fire growth characteristics for smouldering fires, which are based on the work by Quintiere et al. (1982):

$$\dot{m} = a\,t + b\,t^2 \quad 0 \leq t \leq 3600s, \tag{7.2}$$

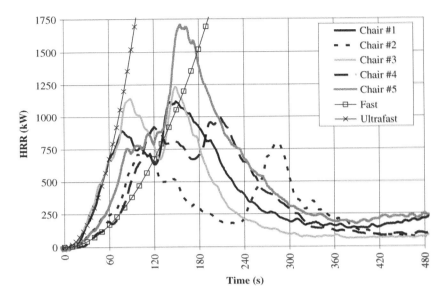

Figure 7.7 Initial heat release rates (HRR) of five chairs of identical design with different fabric coverings are between fast and ultra-fast t-squared fires (from Yung, Wade and Fleischmann, 2004, reproduced by permission of the Society of Fire Safety, Australia).

or

$$\dot{m} = 1.22 \quad 3600s \le t \le 7200s, \tag{7.3}$$

where $\dot{m}$ is the pyrolysis rate in $g \cdot s^{-1}$, $a = 2.78 \times 10^{-5} \, g \cdot s^{-2}$ and $b = 8.56 \times 10^{-8} \, g \cdot s^{-3}$.

7.4.2.2 Fuel Load

Fuel load is the fire load. This governing parameter has influence on whether a fire gets to flashover. It also affects the duration of the fire after flashover (see Figure 7.2). This parameter has no effect on non-flashover fires as the duration of these fires are relatively short. This parameter also has no effect on smouldering fires as these fires are slow burning fires and therefore do not consume a lot of fuel. The duration of post-flashover fires has a direct impact on the structural elements of a building, in addition to posing severe hazards to the occupants. The structural elements of a building are designed to be able to withstand a standard fire in a fire resistance furnace for certain required durations, such as 1 hour, 2 hours and so on. The longer the duration of a post-flashover fire, the more likely is the failure of the structural elements. This will be discussed in more detail in Chapter 8.

Fire loads for specific occupancies can be determined using survey, such as those by Hadjisophocleous and Zalok (2004) and by Bwalya *et al.* (2006). Table 7.4 shows, as an example, the fire loads for various occupancies that are recommended by the British Standards (BSI, 2003, Part 1) for fire safety designs and fire risk assessments.

7.4.2.3 Compartment Geometry and Properties

Compartment geometry and properties refer to the dimensions and shape of the compartment and the thermal properties of the compartment boundaries. They affect the build-up of heat within the compartment as well as the radiant feedback from the compartment boundaries to the fuel. They therefore affect the fire development within the compartment. These parameters can vary depending on the building design. They are also design parameters that are known prior to any fire safety designs or fire risk assessments. They can usually be entered into any comprehensive computer fire model to generate the different design fires for different compartment geometry.

7.4.2.4 Ventilation Conditions

Ventilation includes both natural, such as window and door openings, and mechanical, such as fans. Ventilation depends on the conditions of these windows and doors (open or closed) and fans (on or off). Ventilation conditions affect the supply of fresh air into the compartment and, consequently, the fire development within the compartment. They also affect the outflow of toxic gases which affect the occupants in the building. Ventilation conditions are also design parameters that are known prior to any fire safety designs or fire risk assessments. For example, if doors have self-closers, such as the entrance door of each apartment unit, the doors are probably closed when a fire occurs.

7.4.2.5 Ignition Source, Ignition Location and Fuel Arrangement

These are random input parameters that can not be determined before a fire safety design or fire risk assessment is carried out. As a result of these random parameters, fires can develop into many types, from a small fire to a flashover fire. Instead of considering these random

Table 7.5   Probabilities of fire types for apartment and office buildings with no installed sprinklers (source: Gaskin and Yung, 1997; Eaton, 1989).

Fire type	Apartment buildings			Office buildings		
	Canada (%)	USA (%)	Australia (%)	Canada (%)	USA (%)	Australia (%)
Smouldering fire	19.1	18.7	24.5	–	22.3	21.0
Non-flashover fire	62.6	63.0	60.0	–	53.5	59.1
Flashover fire	18.3	18.3	15.5	–	24.2	19.9

Note: Probabilities for Canadian office buildings were not available due to insufficient fire statistics.

parameters individually, which would mean a large number of possible fire scenarios, an alternative option is to consider the types of fires that have occurred in the past as a result of these random parameters. From fire statistics, three distinct types of fires can be identified, based on the severity of the fire. They are: (1) *smouldering fires* where only smoke is generated, (2) *non-flashover flaming fires* where a small amount of heat and smoke is generated and (3) *flashover fires* where a significant amount of heat and smoke is generated with a potential for fire spread to other parts of the building. Table 7.5 shows, as an example, the probabilities of these three fire types for both apartment and office buildings and for Canada, the United States and Australia. They were obtained based on independent analyses of fire statistics in these three countries (Gaskin and Yung, 1997; Eaton, 1989).

In Table 7.5, the definition of the fire type is based on the severity of the fire when it was observed and recorded by the firefighters upon their arrival. Obviously, small fires can develop into fully developed, post-flashover fires if they are given enough time and the right conditions. For fire safety designs and fire risk assessments, however, the fire conditions at the time of the arrival of the firefighters are the appropriate ones to use. They represent the fire conditions that the occupants are exposed to prior to firefighter extinguishment and rescue operations.

The fire types and their probabilities of occurrence can be used to set up different fire growth scenarios for fire safety designs and fire risk assessments. An example of this is shown in Figure 7.8, where fire scenarios based on the three fire types are shown.

In Figure 7.8, fire initiation is the *probability of fire initiation* in a particular occupancy. The probability of fire initiation can be obtained from fire statistics. Table 7.6 shows, for example, what the British Standards (BSI, 2003, Part 7) recommends for use for apartment and

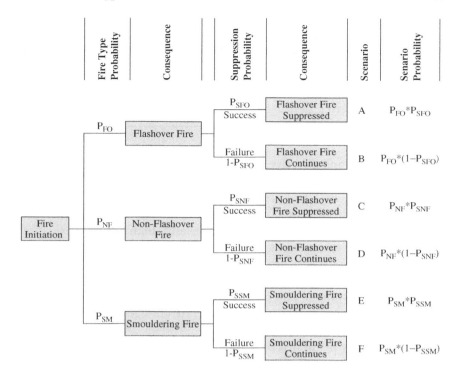

Figure 7.8 Fire growth scenarios based on probabilities of fire types and the success or failure of fire suppression systems.

office buildings. The probability of fire initiation multiplied by the probability of fire type gives the probability of fire initiation of that type of fire.

7.5 Automatic Fire Suppression to Control Fire Growth

One way to control fire development is the use of *automatic suppression* systems, which either extinguish the fire or control it from further development. The probability of success of suppressing a fire depends

Table 7.6 Probability of fire initiations for apartment and office buildings (source: BSI, 2003, Part 7).

Occupancy	Probabilities of fire starts
Apartment buildings	3.0×10^{-3} fires/year/apartment unit
Office buildings	1.2×10^{-5} fires/year/m^2 floor area
Commercial shops	6.6×10^{-5} fires/year/m^2 floor area

on the reliability and effectiveness of the suppression system. Reliability refers to the likelihood the automatic suppression system will activate in the event of a fire, and effectiveness refers to how well the system will extinguish the fire or controls it from further development. The probability of success of suppression is usually high for flashover fires because the heat release rate is high which causes the suppression system, such as *sprinklers*, to activate. The probability of success is not so high for non-flashover fires because the fire may be too small to activate the suppression system. The probability of success is basically zero for smouldering fires because they don't generate enough heat to be able to activate the system.

The probability of success or failure of automatic suppression systems can be used to set up different fire scenarios for fire safety designs and fire risk assessments. A success scenario means the fire is either extinguished or put under control by the fire suppression system. The fire, therefore, does not go anywhere and poses no hazards to the occupants or properties in a building. A failure scenario, on the other hand, means the fire continues to develop as if the fire suppression system does not exist. The fire, therefore, continues to pose hazards to the occupants or properties in a building. An example of this is shown in Figure 7.8, where fire scenarios, based on the success or failure of automatic fire suppressions, are shown.

The probability of suppression can be obtained from fire statistics, although the statistics may show only the suppression probabilities for all fire types, and not for the individual fire types. For example, Table 7.7 shows the sprinklers performance for large fires that should activate the sprinklers (extracted from Kimberly and Hall, 2005, Table 7B). In this table, reliability is the probability of sprinkler activation against large fires that should activate the sprinklers. The fires, therefore, could include both flashover fires and some large non-flashover fires. Effectiveness is the effectiveness of controlling fires once sprinklers are activated. Probability of success is the product of reliability and effectiveness.

Table 7.7   Sprinkler performance (source: Kimberly and Hall, 2005, Table 7-B).

Occupancy	Reliability of activation[a] (%)	Effectiveness of suppression[b] (%)	Probability of success[c] (%)
Apartments	98	98	96
Health care or correctional	96	100	96
One or two family dwelling	94	100	94
Educational	92	100	92
Hotel or Motel	97	94	91
Stores and offices	92	97	90
Manufacturing	93	94	87
Public assembly	90	89	81
Storage	85	90	77

Notes:
[a] Reliability is the probability of sprinkler activation against large fires that should activate. sprinklers.
[b] Effectiveness is the effectiveness of controlling fires once sprinklers are activated.
[c] Probability of Success is the product of reliability and effectiveness.

Sprinkler reliability and effectiveness depend on proper design and maintenance which will be discussed in Chapter 13.

Based on the probabilities of occurrence of the fire types and the probabilities of success or failure of automatic suppression systems, a set of fire growth scenarios can be set up as shown in Figure 7.8. Each scenario has a certain probability of occurrence. For example, Scenario A has a probability of occurrence of $P_{FO}*P_{SFO}$. The use of fire growth scenarios together with the output from fire models under different fire growth scenarios provide the required information that can be used to assess the fire risks to the occupants and properties in a building. For example, fire growth scenarios can be linked up with subsequent fire scenarios such as smoke spread scenarios, occupant evacuation scenarios and so on, to form a complete set of fire scenarios for fire risk assessment, which will be discussed in subsequent chapters. Input to these fire models are the deterministic parameters. Output from these fire models includes flow rate, temperature, smoke, CO and CO_2 concentrations, both within the compartment of fire origin as well as in the exhaust from the compartment.

7.6 Summary

The development of a fire in a compartment is governed by a number of input parameters. These input parameters can be both deterministic and random. Deterministic parameters, which include fuel type, fuel load, compartment geometry and ventilation conditions, are those that can be determined before a fire safety design or fire risk assessment is carried out. Random parameters, which include ignition source, ignition location and fuel arrangement, are those that can not be determined a priori. As a result of these random parameters, fires can develop into many types, from a small fire to a flashover fire.

Instead of considering these random parameters individually, which would mean a large number of possible fire scenarios, an alternative option is to consider the types of fires that have occurred in the past as a result of these random parameters. From fire statistics, three distinct types of fires can be identified, based on the severity of the fire. They are: (1) smouldering fires where only smoke is generated, (2) non-flashover flaming fires where a small amount of heat and smoke is generated and (3) flashover fires where a significant amount of heat and smoke is generated with a potential for fire spread to other parts of the building.

Fire growth scenarios can be set up based on these three fires type. Automatic suppressions can be included in the fire growth scenarios based on the probabilities of success or failure of suppressing the fires. The use of fire growth scenarios, together with the output from fire models under different fire growth scenarios, provides the required information that can be used to assess the fire risks to the occupants and properties in a building. For example, fire growth scenarios can be linked up with subsequent fire scenarios such as smoke spread scenarios, occupant evacuation scenarios and so on, to form a complete set of fire scenarios for fire risk assessment, which will be discussed in subsequent chapters. Input to these fire models are the deterministic parameters. Output from these fire models includes flow rate, temperature, smoke, CO and CO_2 concentrations, both within the compartment of fire origin as well as in the exhaust from the compartment.

7.7 Review Questions

7.7.1 Calculate the probabilities of all the fire scenarios in Figure 7.8. Assume the probabilities of success of suppressing the fires are $P_{SFO} = 0.95$, $P_{SNF} = 0.5$ and $P_{SSM} = 0$. Also, calculate the expected number of flashover fires per year per apartment unit in apartment buildings in the United States that will not be suppressed even with sprinklers installed in these buildings. Assume the sprinklers have the same aforementioned probabilities of success of suppressing the fires. Use Tables 7.5 and 7.6.

7.7.2 If the entrance doors of apartment units have a probability of being open, include the additional fire scenarios in Figure 7.8. There are two additional scenarios: 'door open' and 'door closed', each with a certain probability.

References

Babrauskas, V. (2002) Heat release rates, *SFPE Handbook of Fire Protection Engineering*, 3rd edn, Section 3, Chapter 1, National Fire Protection Association, Quincy, MA.

BSI (2003) Initiation and development of fire within the enclosure of origin (Sub-system 1), *Application of Fire Safety Engineering Principles to the Design of Buildings*, PD7974-1, Part 1, British Standards Institution, London.

BSI (2003) Probabilistic risk assessment, *Application of Fire Safety Engineering Principles to the Design of Buildings*, PD7974-7, Part 7, British Standards Institution, London.

Bwalya, A.C., Carpenter, D.W., Kanabus-Kaminska, M. *et al.* (2006) *Development of A Fuel Package for use in the Fire Performance of Houses Project*, Research Report IRC-RR-207, Institute for Research in Construction, National Research Council Canada, Ottawa, March 2006.

Denize, H. (2000) *The Combustion Behaviour of Upholstered Furniture Materials in New Zealand*, Fire Engineering Research Report 2000/4, University of Canterbury, Christchurch.

Drysdale, D. (1998) *An Introduction to Fire Dynamics*, 2nd edn, John Wiley & Sons, Ltd, Chichester.

Eaton, C. (1989) Fire probabilities – smouldering, flaming or full development? *Fire Safety and Engineering Technical Papers*, Book 1, Part 3, Chapter 7, The Warren Centre, University of Sydney, December 1989, Australia.

Enright, P.A. and Fleischmann, C.M. (1999) CBUF Model I, Applied to Exemplary New Zealand Furniture. Proceedings of the Sixth International Symposium on Fire Safety Science, July 1999, University of Poitiers, France, pp. 147–57.

Gaskin, J. and Yung, D. (1997) *Canadian and U.S.A. Fire Statistics for use in the Risk-Cost Assessment Model*, Internal Report No. 637, Institute for Research in Construction, National Research Council Canada, January 1993, Ottawa.

Girgis, N. (2000) *Full-Scale Compartment Fire Experiments on Upholstered Furniture*, Fire Engineering Research Report 2000/5, University of Canterbury, Christchurch.

Hadjisophocleous, G. and Zalok, E. (2004) Development of Design Fires for Commercial Buildings. Proceedings of the International Conference on Fire Safety Engineering, March 2004, Sydney.

ISO (1993) *International Standards – Fire Tests – Full Scale Room Test for Surface Products*, ISO9705:1993(E), International Organization for Standardization, Geneva.

ISO (2003) Fire Safety Engineering, Part 2 – Design fire Scenarios and Design fires, ISO/TC92/SC4/WG6/N49 draft document, International Organization for Standardization, Kyoto meeting, Japan, April 2003.

Karlsson, B. and Quintiere, J.G. (2000) *Enclosure Fire Dynamics*, CRC Press LLC, Boca Raton, FL.

Kimberly, D.R. and Hall, J.R., Jr. (2005) *U.S. Experience with Sprinklers and other Fire Extinguishing Equipment*, NFPA Report, August 2005, Table 7B: Sprinklers Performance, National Fire Protection Association, Quincy, MA.

McGrattan, K.B. and Forney, G. (2006) *Fire Dynamics Simulator (Version 4): user's Guide*, NIST Special Publication 1019, National Institute of Standards & Technology, March 2006, Gaithersburg, MD.

Peacock, R.D., Jones, W.W., Reneke, P.A. and Forney, G.P. (2005) *CFAST: Consolidated Model of Fire Growth and Smoke Transport (Version 6): user's Guide*, NIST Special Publication 1041, National Institute of Standards and Technology, August 2005, Gaithersburg, MD.

Quintiere, J.G., Birky, M., Macdonald, F. and Smith, G. (1982) An analysis of smouldering fires in closed compartments and their hazard due to carbon monoxide. *Fire and Materials*, 6(3-4), 99–110.

SFPE (2002) Hazard calculations, *SFPE Handbook of Fire Protection Engineering*, 3rd edn, Section 3, National Fire Protection Association, Quincy, MA.

Tewarson, A. (2002) Generation of heat and chemical compounds in fires, *SFPE Handbook of Fire Protection Engineering*, 3rd edn, Section 3: Chapter 4, National Fire Protection Association, Quincy, MA.

Wade, C.A. (2002) BRANZFIRE Technical Reference Guide, BRANZ Study Report 92, Building Research Association of New Zealand, New Zealand.

Yung, D. and Benichou, N. (1999) Design Fires for Fire Risk Assessment and Fire Safety Designs. Proceedings of the 4th Fire Risk and Hazard Research Application Symposium, June 23–25, 1999 San Diego, CA, pp. 101–12.

Yung, D. and Benichou, N. (2002) How design fires can be used in fire hazard analysis. *Fire Technology*, 38(3), 231–42.

Yung, D., Wade, C. and Fleischmann, C. (2004) Development of Design Fires for Buildings. Proceedings of the International Conference on Fire Safety Engineering, March 2004, Sydney.

8

Fire Spread Probabilities

8.1 Overview

This chapter discusses the probability of fire spread in a building as a result of the failure of the boundary elements. The boundary element would fail if its fire resistance rating (FRR) is not high enough to withstand a fully developed compartment fire. The level of fire resistance that a building element can provide is measured in a standard-fire resistance test employing a fire furnace and a controlled standard fire. In the real world, however, fire development in a compartment does not necessarily follow that of a standard fire. Methods have been developed to equate the severity of real-world fires to equivalent standard fires. This allows a building component with a certain FRR to be assessed for fire resistance failure against any real-world fire. Also, if the real-world fires have a probability distribution, then the equivalent standard fires also have a probability distribution. The probability of failure can be calculated based on the magnitude of the FRR against the mean and standard deviation of the probability distribution of the standard fires.

Fire spread across boundary elements from compartment to compartment can take many paths. The probability of fire spread of each path depends on the probabilities of developing into fully developed fires in all the compartments and the probabilities of failure of all the boundary elements that are involved in each path. The combined probability of all the probable fire spread paths from one compartment to another is the overall probability of fire spread. Fire spread is a relatively slow process because of the relatively long time it takes to fail each boundary element. The calculation of the probability of fire spread is usually for the non-time-dependent assessment of the probability of property damage and less for occupant safety. It has implications for occupant safety only when occupants are trapped in certain compartments and the emergency responders cannot get to them quickly.

8.2 Fire Resistant Construction

To prevent *structural collapse* and *fire spread* in buildings, as a result of fires, buildings are normally required by building regulations to be constructed such that their building components can provide a certain level of fire resistance. Fire resistance is the ability of a building component, such as a column or a wall, to resist a fire without *structural failure* or without allowing the fire to pass through the building component. Usually, the taller or bigger the building is, the higher are the levels of fire resistance that are required of its building components.

The level of fire resistance that a building component can provide is measured in a *standard-fire resistance test* employing a fire furnace and a controlled fire called *standard fire* (details about the standard test will be described later in Section 8.2.1). In such a standard-fire resistance test, the measured duration, in minutes or hours, that a building component can withstand a standard fire is called the fire resistance rating *(FRR)*. In building regulations, the levels of fire resistance that are required of different building components are regulated through the specifications of their FRRs. Based on the structural as well as the fire safety importance of the various building components, different FRRs are required for different building components. For example, major structural components, such as columns and beams, are usually required to have a high FRR of between 2 and 4 hours. The stringent requirement provides an ample time for the occupants to evacuate the building and the firefighters to control the fire without the worry of a structural collapse. The non-structural components, such as partition walls and doors, are usually required to have a lower FRR of between 1 and 2 hours. This lower requirement still provides an adequate time for the occupants to evacuate the building and the firefighters to combat the fire without the worry of a fire spread from the compartment of fire origin to other locations in the building.

It should be noted that high-rise buildings are seldom collapsed by fires because of the high FRRs normally required of their structural components. The collapse of the twin towers in the World Trade Center on September 11, 2001 was a rare event. Comprehensive studies led by the National Institute of Standards and Technology in the United

States have concluded that fire alone was not likely to have led to the collapse of the twin towers (NIST, 2005). Other unusual factors, such as the high-speed impact damage to the principal structural members, the dislodging of the fireproof insulation and the disabling of the water supplies to the automatic sprinkler systems, as a result of the impact of the two high-speed aircrafts, contributed to the collapse.

Even with a weakened structure and loss of fire protection, Towers 1 and 2 in the World Trade Center withstood the fire for 102 minutes and 56 minutes respectively before they collapsed. The relatively long time before collapse allowed 99 % of those below the impact floors to evacuate successfully. A total of 2749 people perished in that incident, including emergency responders and those in the two aircrafts, but not counting the 10 hijackers. The NIST study concludes with eight major groups of recommendations: increased structural integrity, enhanced fire endurance, new methods for fire resistant design of structures, improved active fire protection, improved building evacuation, improved emergency response, improved procedures and practices, and improved education and training. The essence of these recommendations is to help minimize the probability of building collapse and to expedite occupant evacuation and emergency response. (Review Chapter 2 on five fire barriers to minimize fire risk.)

8.2.1 Fire Resistance Test

Standard fire resistance tests are conducted in large fire furnaces. A full-scale specimen, such as a wall or a floor, is mounted on one side of the furnace with one side of the specimen facing the inside of the furnace and the other facing the outside. The inside face of the specimen is then exposed to a fire that is created inside the furnace using natural gas, propane or oil. In the case of a fire resistance test for structural columns, the whole column specimen is placed inside a column furnace. Hydraulic loading can be applied to the specimen during the test to simulate actual structural loading on the specimen. Figure 8.1 shows the view of a steel column seen through an opened door of a full-scale column furnace at the end of a typical fire resistance test.

In a standard-fire resistance test, the fire inside the furnace is controlled to simulate a certain fire development which has been assumed by a standard organization to represent a typical fire development. Basically, the temperature inside the furnace is controlled to follow a certain temperature rise as a function of time. Such prescribed time-temperature curve is called a *standard fire*. There are different standard fires that

Figure 8.1 A steel column seen through an opened door of a column fire furnace at the end of a fire resistance test (photo courtesy of Dr Venkatesh Kodur, reproduced by permission of the National Research Council Canada).

have been developed by different organizations for fire resistance tests (Buchanan, 2001). The most widely used ones are the *ASTM E119* (ASTM, 1998) and the *ISO 834* (ISO, 1975) standard fires. There is also the *hydrocarbon standard fire* (EC1, 1994) that simulates a pool fire of a liquid hydrocarbon fuel. Figure 8.2 shows the time-temperature curves of the three standard fires up to 180 minutes. This figure shows that the ASTM E119 and the ISO 834 standard fires follow each other very closely and that the hydrocarbon standard fire is a more severe fire. The ASTM E119 and the ISO 834 standard fires rise rapidly in the first 10 minutes to about 700 °C and then rise slowly in the next 170 minutes to about 1100 °C. The hydrocarbon standard fire rises much more rapidly in the early stage of the fire development, reaching its maximum value of 1100 °C in about 20 minutes.

In fire resistance tests, the FRR of a building element is measured by the time duration that a test specimen can withstand a standard fire before failure. Failure is assumed when the test specimen arrives at any one of the failure conditions that are considered precursors to either structural collapse or fire spread. Three failure conditions are used to

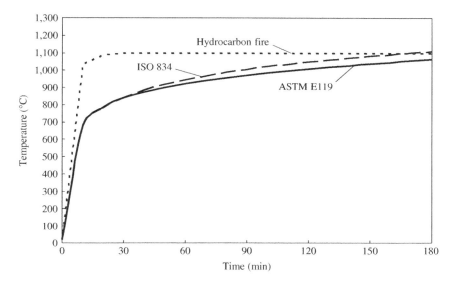

Figure 8.2 Standard fires (prescribed time-temperature curves) for use in fire resistance tests.

determine failure: stability failure, integrity failure and insulation failure. Stability failure is the failure condition when the test specimen can not maintain its structural stability and begins to collapse. Integrity failure is the failure condition when the test specimen can not maintain its integrity and allows small flames or smoke to pass through. The small flames and smoke have the potential of igniting combustible materials that are close to the unexposed side of the building component and hence can cause fire spread. Insulation failure is the failure condition when the unexposed side of the test specimen has an average temperature rise of more than 140 °C, or a maximum temperature rise at any single location of more than 180 °C. The high temperatures have the potential, similar to small flames and smoke, of igniting combustible materials that are close to the unexposed side of the building component and hence can cause fire spread. It should be noted that fire resistance tests are not the major focus of this book; whereas the probability of fire spread is. For more details on fire resistance tests, consult the book *Structural Design for Fire Safety* (Buchanan, 2001).

Full-scale fire resistance tests are generally very expensive to conduct. Engineering methods are continually being developed by the fire research community to help predict FRRs of building components. Many of these methods for steel, concrete and timber components can be found in the SFPE Handbook (SFPE, 2002b).

8.3 Probability of Failure

Fire resistance rating is a simple rating system that gives only an indication of the level of fire resistance that a building component can provide. It only gives the level of fire resistance that a building component can provide against a standard fire. It does not give the level of fire resistance that a building component can provide against any fire. For example, if a building component has a FRR of 60 minutes, it means the building component can resist a standard fire that lasts less than 60 minutes. It does not imply that the building component can resist any fire that lasts less than 60 minutes. In the real world, however, fires do not necessarily follow the standard-fire time-temperature curve. As was discussed in Chapter 7, fire growth follows different paths depending on the parameters that govern its growth and decay. If *real-world fires* do not follow the standard-fire time–temperature curve, there is a need to include real-world fires in the fire resistance regulations of building components (Almand, 2006).

One way to include a real-world fire in the evaluation of FRRs of building components is to find a way to equate the severity of a real-world fire to that of a standard fire (Buchanan, 2001). As fire severity can be measured by the maximum temperature rise in a building component, a real-world fire can be equated to a standard fire that achieves the same temperature rise in the building component as the real-world fire does. For example, if a building component, such as a protected steel beam, has a maximum temperature rise of 400 °C as a result of exposure to a real-world fire and if it takes a standard fire 50 minutes to achieve the same temperature rise in the building component, then the real-world fire is considered to be equivalent to a 50-minutes standard fire. This allows a building component with a certain FRR to be assessed for fire resistance failure against any fire.

There are many ways to equate the severity of a real-world fire to that of a standard fire. One method that is widely used is the one published by *CIB W14* (International Council for Research and Innovation in Building and Construction) which is described in the book Structural Design for Fire Safety (Buchanan, 2001). The CIB W14 method equates the severity of a real-world fire to that of an ISO 834 standard fire, based on three of the parameters that govern fire development (review Chapter 7). The three parameters that CIB W14 uses are the fuel load density, the compartment boundary parameter and the ventilation factor, as described in the following equation:

$$t_e = e_f \, k_c \, w, \tag{8.1}$$

where t_e is the equivalent time (minute) of the ISO 834 standard fire, e_f is the fuel load density (MJ·m^{-2}), k_c is the compartment boundary parameter (min·m$^{2.25}$·MJ^{-1}), and w is the ventilation factor (m$^{-0.25}$).

The compartment boundary parameter k_c is a parameter that depends on the thermal inertia $\sqrt{k\rho c_p}$ of the boundary, where k is the thermal conductivity (W·m^{-1}·K^{-1}), ρ is the density (kg·m^{-3}) and c_p is the specific heat (J·kg^{-1}·K^{-1}). Table 8.1 shows that the equivalent time of the standard fire is lower when the thermal inertia is higher. Nominal thermophysical properties of some common construction materials are shown in Table 8.2. This table shows that wood boundaries have low thermal inertia, steel boundaries have high thermal inertia and gypsum, brick and concrete boundaries have intermediate thermal inertia.

The ventilation parameter w is given by the following equation:

$$w = \frac{A_f}{\sqrt{A_v A_t \sqrt{H_v}}}, \tag{8.2}$$

where A_f is the floor area of the compartment (m^2), A_t is the total area of internal surfaces (m^2), A_v is the total area of openings in the walls (m^2) and H_v is the height of the openings (m).

Table 8.1   Dependence of the compartment boundary parameter k_c on the thermal inertia $\sqrt{k \cdot \rho \cdot c_p}$ (source: Buchanan, 2001).

$\sqrt{k \cdot \rho \cdot c_p}$ (J·K^{-1}·m^{-2}·s$^{-0.5}$)	k_c (min·m$^{2.25}$·MJ^{-1})
<720	0.09
720–2500	0.07
>2500	0.05

Table 8.2   Nominal thermophysical properties of some common construction materials (source: SFPE, 2002a).

Material	K (W·m^{-1}·K^{-1})	ρ (kg·m^{-3})	c_p (J·kg^{-1}·K^{-1})	$\sqrt{k \cdot \rho \cdot c_p}$ (J·K^{-1}·m^{-2}·s$^{-0.5}$)
Wood (fir)	0.11	420	2700	350
Gypsum	0.48	1400	840	750
Brick	0.69	1600	840	960
Concrete	1.40	2100	880	1610
Carbon steel ($C \approx 1.0$ %)	43.0	7800	470	12 600

We will use an example to demonstrate how the CIB W14 method works. We will use an apartment unit in an apartment building to work out what the equivalent ISO 834 standard fire is. Let us assume that the apartment unit is 12.0 m wide, 8.0 m deep and has a ceiling height of 2.5 m. Total window opening is 6.0 m wide by 1.5 m high. Boundary walls and floors are of concrete construction. Fuel load density is that of the mean (50-percentile) value of the British Standards, 780 MJ m^{-2} (see Table 7.4, Chapter 7). Using Equations 8.1 and 8.2 and Tables 8.1 and 8.2, the fire can be worked out to be equivalent to an ISO 834 standard fire of 92 minutes. The 92-minutes fire represents the mean fire; which means 50 % of the equivalent standard fires are less than 92 minutes and the other 50 % are longer than 92 minutes. If the fuel load density is assumed to be that of the 90-percentile value of the British Standards, 920 MJ m^{-2} (see Table 7.4, Chapter 7), then the fire is worked out to be equivalent to an ISO 834 standard fire of 109 minutes. The 109-minutes fire represents the 90-percentile fire; which means 90 % of the equivalent standard fires are less than 109 minutes and the other 10 % are longer than 109 minutes.

The above example shows that if the fuel load density has cumulative probabilities associated with its different values, then the corresponding equivalent standard fire also has cumulative probabilities associated with its different values. And if the corresponding equivalent standard fire has cumulative probabilities associated with its different values, then the higher the FRR of the boundary walls and floors is, the higher is the percentage of the probable fires that can be resisted. For example, based on the above example, if the apartment boundary walls and floors have a FRR of 92 minutes, they can resist 50 % of the probable fires. That is, the probability of success is 50 % and the probability of failure is also 50 %. If they have a FRR of 109 minutes, they can resist 90 % of the probable fires. The probability of success is now 90 % and the probability of failure is only 10 %.

In the CIB W14 method, the severity of a fire is shown to depend on three governing parameters: the fuel load density e_f, the compartment boundary parameter k_c, and the ventilation factor w. Of these three governing parameters, two are related to the building design and are, therefore, easily determined. They are the compartment boundary parameter k_c and the ventilation factor w. The one parameter that cannot be easily determined is the fuel load density e_f. Its value is usually obtained by survey and can vary a lot from country to country. For example, for apartment units, the British Standards recommend a mean fuel load density of 780 MJ m^{-2} (see Table 7.4, Chapter 7), whereas

Canadian researchers at the National Research Council Canada found a much lower value of $350\,\mathrm{MJ\,m^{-2}}$ for Canadian living areas (see Section 7.4.1, Chapter 7). In addition, the fuel load density usually has a probability distribution of different values (a range of values with each value having a certain probability of occurring). Because the fuel load density has this *probability distribution* of different values, the translation of the probable fires into equivalent standard fires would also have a probability distribution of different equivalent standard fires.

Figure 8.3 shows a probability distribution of equivalent standard fires t, with a mean value μ, and a standard deviation σ. Figure 8.3 also shows a vertical line representing the FRR of the boundary elements (building components that form the enclosed space). The FRR is assumed here to have a fixed value without a probability distribution. If the FRR has a probability distribution, the problem becomes a more complex one which will be discussed in Chapter 12. In Figure 8.3, the area under the curve and to the left of the FRR represents all the probable fires that can be resisted by the boundary elements with this FRR. The shaded area, on the other side of the FRR line, represents all the probable fires that can not be resisted by the boundary elements with this FRR. The shaded

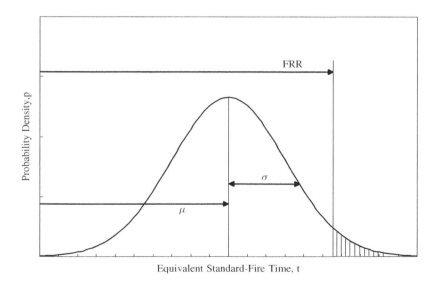

Figure 8.3 Probability distribution of equivalent standard fires t, with a mean value μ, and a standard deviation σ. The area under the curve and to the left of the fire resistance rating (FRR) represents all the probable fires that can be resisted by the FRR. The shaded area to the right of the FRR represents all the probable fires that can fail the FRR. The shaded area therefore represents the probability of failure.

area, therefore, represents the probability of failure. As can be easily seen from this figure, the probability of failure depends on how far the FRR is away from the mean μ and the standard deviation σ. The farther the FRR is away from the mean, the smaller is the shaded area. Also, the narrower the standard deviation is, the smaller is the shaded area.

The shaded area in Figure 8.3, representing the probability of failure, is obtained in general by mapping of the real-world fires into equivalent standard fires and then by measurement of the shaded area. However, if the probability distribution follows one of the well-known mathematical functions, then the probability distribution can be mathematically integrated to give the probability of failure. For example, if the probability distribution of the fuel load density follows that of a *normal distribution*, then the equivalent standard fire would also have a probability distribution that follows a normal distribution. If this distribution has a mean value, μ, and a standard deviation σ, as shown in Figure 8.3, then the integration of the probability distribution function from $-\infty$ to the FRR gives the *cumulative probability P(FRR)* of all the probable equivalent standard fires that are less severe than the FRR. Equation 8.3 shows that the integration results in a cumulative standard normal distribution function Φ (Wolfram MathWorld, 2007) with $P(FRR)$ as a function of the non-dimensional parameter $(FRR - \mu)/\sigma$.

$$P(FRR) = \frac{1}{\sqrt{2\pi}\sigma} \int_{-\infty}^{FRR} e^{-\frac{(t-\mu)^2}{2\sigma^2}} \, dt = \Phi\left(\frac{FRR - \mu}{\sigma}\right). \qquad (8.3)$$

Note that the integration in Equation 8.3 is from $-\infty$ to the FRR and not from 0 to the FRR. It should be integrated from 0 to the FRR because the equivalent standard fire has no negative values. However, because the integrand has a value that is practically 0 when the variable t is less than 0, the integration is the same whether it is integrated from $-\infty$ to the FRR or from 0 to the FRR. By changing the integration from $-\infty$ to the FRR, the integration becomes that of a cumulative standard normal distribution Φ.

If $P(FRR)$ is the cumulative probability of all the probable fires up to the FRR, then the complement, $P'(FRR) = 1 - P(FRR)$, is the cumulative probability of all the probable fires that are more severe than the FRR. Equation 8.4 shows that the complement cumulative probability $P'(FRR)$, which by definition is the probability of failure, is equal to $\Phi(-(FRR - \mu)/\sigma)$.

$$P'(FRR) = 1 - P(FRR) = \Phi\left(-\frac{FRR - \mu}{\sigma}\right). \qquad (8.4)$$

Equation 8.4 shows that the probability of failure is a function of the non-dimensional parameter $(FRR - \mu)/\sigma$, which is a measure of how far the FRR is away from the mean μ and the standard deviation σ of the probability distribution of equivalent standard fires. The larger this non-dimensional parameter is, the farther is the FRR away from μ and σ. Thus, the larger the non-dimensional parameter is, the smaller is the shaded area, and the smaller is the probability of failure. This relationship is plotted in Figure 8.4 where the probability of failure is shown, as expected, to decrease with the increase of the non-dimensional parameter $(FRR - \mu)/\sigma$. At $(FRR - \mu)/\sigma = 0$, the probability of failure is shown to be equal to 0.5, or 50 %. This is the case because when $FRR = \mu$, the boundary element has a FRR that can resist 50 % of the probable fires. At $(FRR - \mu)/\sigma = 2$, the probability of failure is shown to be equal to approximately 0.02, or 2 %. This is again expected because when the difference between the FRR and the mean μ is equal to two times the standard deviation σ the boundary element has a FRR that covers most of the probable fires under the probability distribution curve (see Figure 8.3).

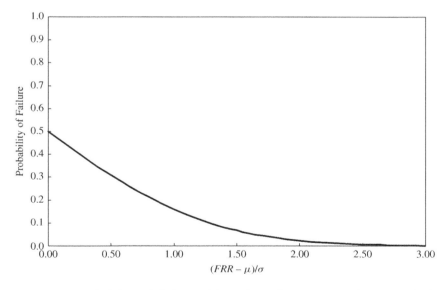

Figure 8.4 Probability of failure of boundary elements as a function of the non-dimensional parameter $(FRR - \mu)/\sigma$, which is a measure of how far the fire resistance rating (FRR) is away from the mean μ and the standard deviation σ. FRR is the fire resistance rating of the boundary elements; μ and σ are the mean and standard deviation, respectively, of the probability distribution of the equivalent standard fires (see Figure 8.3).

We will use an example to show how Equation 8.4 and Figure 8.4 can be used to calculate the probability of failure. We will use the previous example, based on the fuel load density values recommended by the British Standards for dwellings, to calculate what the probability of failure is if the FRR is 120 minutes. In the previous example, the probability distribution of the fuel load density is translated, using the CIB W14 method, to a probability distribution of equivalent standard fires with a mean μ of 92 minutes and a 90-percentile value of 109 minutes. Given these values, the normal distribution is defined. If the FRR that covers 90 % of the probable fires is 109 minutes, then the non-dimensional parameter $(FRR - \mu)/\sigma$ has a value of 1.28, obtained from Equation 8.3. (Inserting this value in Equation 8.3 gives a cumulative probability $P(FRR)$ of 90 %.) If $(FRR - \mu)/\sigma = 1.28$, $FRR = 109$ minutes and $\mu = 92$ minutes, then σ can be calculated to have a value of 13 minutes. With the standard deviation σ and the mean μ defined, we can calculate the probability of failure for any FRR. For example, if the FRR is 120 minutes, the non-dimensional parameter $(FRR - \mu)/\sigma$ has a value of 2.11. From Figure 8.4, or more directly from Equation 8.4, the probability of failure is 1.7 %.

8.4 Fire Spread Probabilities

8.4.1 Fire Spread across One Boundary Element

A compartment is a space that is enclosed by *boundary elements*, such as walls, floors, ceilings, doors and windows. If a door between two compartments is open, then the two compartments form one single fire compartment. The probability of fire spread across a boundary element from one compartment to an adjacent one depends on the probability of failure of the boundary element that separates them. In addition, a fire needs to develop into a fully developed fire first within a compartment before it has the potential to fail the boundary elements. In Chapter 7, the probability of an ignition developing into a fully developed fire depends on the probability of ignition, the probability of developing into a flashover fire, and the probability of not suppressing the flashover fire by a sprinkler system if it is installed, as described by the following Equation 8.5:

$$P_{FD} = P_{IG} \times P_{FO} \times (1 - P_{SFO}), \tag{8.5}$$

where P_{FD} is the probability of an ignition developing into a fully developed fire, P_{IG} is the probability of ignition, P_{FO} is the probability of developing into a flashover fire, and P_{SFO} is the probability of suppressing the flashover fire by a sprinkler system. The probability of fire spread P_{FS} across one boundary element, therefore, depends on the probability of an ignition developing into a fully developed fire in a compartment and the probability of failure of the boundary element, as described by the following Equation 8.6:

$$P_{FS} = P_{FD} \times P'(FRR). \tag{8.6}$$

Equations 8.5 and 8.6 govern the probability of fire spread across one boundary element. In the compartment of fire origin, the probability of ignition P_{IG} is a result of random events. It is normally obtained from fire statistics, as was discussed in Chapter 7. The probability of ignition in the second compartment, on the other hand, is a result of the failure of the boundary element between the compartment of fire origin and the second compartment. The probability of ignition in the second compartment, therefore, is equal to the probability of fire spread P_{FS} across the boundary element from the compartment of fire origin to the second compartment. Similarly, the probability of ignition in subsequent compartments is equal to the probability of fire spread across the boundary element from the preceding compartment to the next compartment.

8.4.2 Fire Spread across Multiple Boundary Elements

In a building, the spread of a fire crossing boundary elements from one compartment to another can take many paths and each path can cross many boundary elements. Fire spread crossing boundary elements, therefore, is a relatively slow process because of the long time it takes to fail the boundary elements – one at a time along the path. This is a much slower process than the spread of a fire within a compartment which can be very quick, as was discussed in Chapter 7. This is also a much slower process than the spread of smoke in a building which can be even quicker, as will be discussed in Chapter 9. Fire spread within a compartment poses significant risks to the occupants in the compartment because of the quickness of the spread. Similarly, smoke spread poses even greater risks to the occupants in a building because of the even faster speed of the spread. Fire spread across boundary elements from compartment to compartment, on the other hand, is relatively slow. The time dependence of the spread, which has impact on occupant safety, is normally not considered. It is considered mainly for the non-time-dependent assessment of the probability of property damage. It has implications for occupant safety only when occupants are trapped in certain compartments and the emergency responders cannot get to them before the fire spread gets to them.

The probability of fire spread from one compartment to another is based on all the probable paths that the fire spread can take. The combined probability of all probable paths is the overall probability of fire spread from one compartment to another. We will use a simple example to show how this works. Figure 8.5 shows an example of a network diagram of a three-storey building that can be used to identify the paths and calculate the probabilities of fire spread (Benichou, Yung and Dutcher, 2001). In this diagram, circles denote the compartments, corridors, stairwells, elevator shafts and ducts; arrows denote boundary elements and the directions of the arrows denote the directions of the fire spread paths. Also in this simple example, all the compartments on one floor are combined into one, and all the elevator shafts, stairwells and ducts are also combined into one elevator shaft, one stairwell and one duct, respectively. This implies that the combined enclosed spaces are considered identical in terms of the probability of fire spread through them. In addition, the stairwell is considered to have some fire resistance from floor to floor because there are usually no combustibles in the stairwell and any fire that breaches the stairwell wall or door does not automatically spread to the whole stairwell. The elevator shafts and ducts, on the other hand, are considered to have some combustibles in

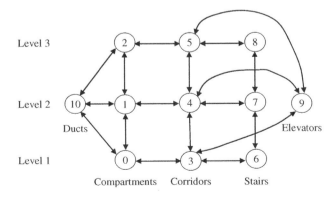

Figure 8.5 An example of a network diagram of a three-storey building with circles denoting the compartments, corridors, stairs, elevators and ducts; arrows denoting boundary elements and the directions of the arrows denoting the directions of the fire spread paths.

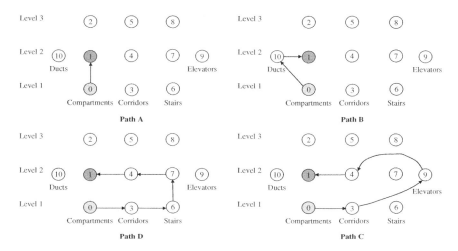

Figure 8.6 An example of four different fire spread paths from compartment 0 to compartment 1, with path A as a direct path, path B involving the ducts, path C involving the elevators and path D involving the stairs.

them and any fire that breaches them would spread to the whole duct or shaft.

For illustration purposes only, we will consider only four different fire spread paths from compartment 0 to compartment 1, as shown in Figure 8.6. Path A is a direct path through the ceiling of compartment 0 to compartment 1; path B involves the ducts; path C involves the elevators; and path D involves the stairs. The probability of each path is based on the probabilities of fire spread across all the boundary elements associated with each path, as described by Equation 8.7. The combined probability of multiple paths from compartment 0 to compartment 1 is given by Equation 8.8, which is formulated based on combined probabilities of non-mutually exclusive events. As was discussed in Section 8.3, the probability of failure of each boundary element $P'(FRR)$ is usually different, depending on the severity of the fire and the FRR of the boundary element. Also, the probability of an ignition developing into a fully developed fire P_{FD} in each compartment may be different. If we assume, for simplicity, that all boundary elements have the same probability of fire spread P_{FS} of 0.3, the probabilities of fire spread of single paths and the combined probabilities of multiple paths are shown in Table 8.3. This table shows that as the number of boundary elements in a single path increases, the probability of fire spread decreases and its contribution to the overall probability of multiple paths from one

Table 8.3 Probabilities of fire spread for single paths and combined probabilities for multiple paths, assuming all boundary elements have the same probability of fire spread of 0.3.

Single path	Probability of a single path	Combined paths	Probability of combined paths
A	0.3000	A	0.3000
B	0.0900	A + B	0.3630
C	0.0081	A + B + C	0.3682
D	0.0024	A + B + C + D	0.3697

compartment to another diminishes.

$$P_A = P_{0-1}$$
$$P_B = P_{0-10} \times P_{10-1}$$
$$P_C = P_{0-3} \times P_{3-9} \times P_{9-4} \times P_{4-1}$$
$$P_D = P_{0-3} \times P_{3-6} \times P_{6-7} \times P_{7-4} \times P_{4-1}$$
(8.7)

$$P_{A+B} = P_A + P_B - (P_A \times P_B)$$
$$P_{A+B+C} = P_{A+B} + P_C - (P_{A+B} \times P_C)$$
$$P_{A+B+C+D} = P_{A+B+C} + P_D - (P_{A+B+C} \times P_D)$$
(8.8)

It should be noted that the above example is a simplified one and is for illustration purposes only. In high-rise buildings with many compartments, the calculation of fire spread probabilities is usually more complex. Not only are there more compartments and more boundary elements to be considered, there are also more potential paths of fire spread to be considered. Each compartment may have a different probability of developing into fully developed fires and each boundary element may have a different probability of failure. In such cases, a computer program is needed to search for all probable paths and calculate the overall probability from one compartment to another. Benichou *et al.* at the National Research Council Canada have found an efficient algorithm to do just that (Benichou, Yung and Dutcher, 2001).

8.5 Summary

The level of fire resistance that a building component can provide is measured in a standard fire resistance test employing a fire furnace and a controlled standard fire. In the real world, however, fire development in a compartment does not necessarily follow that of a standard fire. Methods have been developed to equate the severity of real-world fires to equivalent standard fires. This allows a building component with a certain FRR to be assessed for fire resistance failure against any real-world fire. Also, if the real-world fires have a probability distribution, then the equivalent standard fires also have a probability distribution. The probability of failure can be calculated based on the magnitude of the FRR against the mean and standard deviation of the probability distribution of the standard fires.

Fire spread across boundary elements from compartment to compartment can take many paths. The probability of fire spread of each path depends on the probabilities of developing into fully developed fires in all the compartments and the probabilities of failure of all the boundary elements that are involved in each path. The combined probability of all the probable fire spread paths from one compartment to another is the overall probability of fire spread. Fire spread is a relatively slow process because of the relatively long time it takes to fail each boundary element. The calculation of the probability of fire spread is usually for the non-time-dependent assessment of the probability of property damage and less for occupant safety. It has implications for occupant safety only when occupants are trapped in certain compartments and the emergency responders cannot get to them quickly.

8.6 Review Questions

8.6.1 Calculate the probability of fire resistance failure for an apartment unit with concrete construction and a FRR of 60 minutes. Assume that the apartment unit is 12.0 m wide, 8.0 m deep, has a ceiling height of 2.5 m, and a total window opening of 6.0 m wide by 1.5 m high. Assume that the fuel load density has a mean value of 350 MJ m^{-2} and a standard deviation of 105 MJ m^{-2}. Review the example in the text.

8.6.2 Calculate the probability of fire spread in a three-storey building, the same as described in Figure 8.5, from an apartment on level 1 (compartment 0) to an apartment on level 3 (compartment 2). Assume all the boundary elements have the same probability of fire spread as that which is calculated in the above question 8.6.1. Use all the probable paths to find the overall probability of fire spread. Paths with a large number of boundary elements have little contribution to the overall probability of fire spread.

References

Almand, K. (2006) *Improving the Technical Basis for Testing and Design of Fire-Resistant Structures, Emerging Trends eNewsletter*, Society of Fire Protection Engineers, Issue 3, http://www.fpemag.com/archives/enewsletter.asp?i=5

ASTM (1998) *Standard Test Methods for Fire Tests of Building Construction and Materials*, ASTM E119, American Society for Testing and Materials, Philadelphia.

Benichou, N., Yung, D. and Dutcher, C. (2001) A Model for Calculating the Probabilities of Flame Spread from Fires in Multi-Storey Buildings. Proceedings of the International Conference on Building Envelope Systems and Technologies (ICBEST), Ottawa, Canada, June 2001, pp. 407–11.

Buchanan, A.H. (2001) 'Fire Severity' and 'Fire Resistance', in *Structural Design for Fire Safety*, Chapters 5 and 6, John Wiley & Sons Ltd, England.

EC1 (1994) *Actions on Structures Exposed to Fire, Eurocode 1: Basis of Design and Design Actions on Structures*, ENV 1991-2-2, European Committee on Standardization, Brussels, Belgium.

ISO (1975) *Fire Resistance Tests – Elements of Building Construction*, ISO 834, International Organization for Standardization, Geneva, Switzerland.

NIST (2005) *Federal Building and Fire Safety Investigation of the World Trade Center Disaster: Final Report on the Collapse of the World Trade Center Towers*, NIST NCSTAR 1, National Institute of Standards and Technology, Gaithersburg.

SFPE (2002a) Thermophysical property data, in *SFPE Handbook of Fire Protection Engineering 2002*, 3rd edn, Appendix B, National Fire Protection Association, Quincy, MA.

SFPE (2002b) Analytical methods for determining fire resistance of steel, concrete and timber members, in *SFPE Handbook of Fire Protection Engineering 2002*, 3rd edn, Sec. 4, Chapters 9–11, National Fire Protection Association, Quincy, MA.

Wolfram MathWorld (2007) Normal Distribution, http://mathworld.wolfram.com/NormalDistribution.html

9

Smoke Spread Scenarios

9.1 Overview

A fire's development in a compartment may not only cause physical harm to the occupants and properties in the compartment in which the fire originates, but also be a risk to the occupants and properties in other parts of the building. Heat, toxic gases and smoke from the compartment in which the fire originates can spread rapidly to other locations in a building, and these can pose life risks to the occupants and financial risks to the properties in these other locations. *Smoke spread* is the common term for the spread of heat, toxic gases and smoke in a building, and refers to the spread of toxic gases and heat as well as smoke particles.

Smoke spread in a building is governed by:

- buoyancy force
- stack effect
- wind effect

Calculating smoke spread is usually performed by the use of computer smoke spread models. To prevent smoke spread in order to minimize risks to occupants and properties, smoke control systems are normally required in building regulations. There are basically three types of smoke control strategies:

1. door self-closers and automatic shut-offs of mechanical ventilation,

Principles of Fire Risk Assessment in Buildings   D. Yung
© 2008 John Wiley & Sons, Ltd

2. automatic fire floor smoke extraction and pressurization of the floors above and below and
3. stairwell and elevator shaft pressurization.

Smoke spread scenarios can be constructed based on the success or failure of these control systems. The probabilities of smoke spread scenarios and the time-dependent values of the smoke spread parameters of *temperature*, CO, CO_2 and *soot concentration*, may be used to assess the life risks to the occupants and financial risks to the properties.

9.2 Smoke Spread Characteristics and Modelling

As was described in previous chapters, the development of a fire in a compartment poses not only physical harms to the occupants and properties in the compartment of fire origin, but also risks to the occupants and properties in the other locations in a building. The outflow of heat, *toxic gases* and smoke from the compartment of fire origin can spread quickly to the other locations in a building, posing life risks to the occupants and financial risks to the properties in these other locations. The spread of heat, toxic gases and smoke in a building is commonly referred to as *smoke spread*. Smoke spread, therefore, refers to the spread of not only smoke particles but also toxic gases and heat.

The physical parameters that govern the development of a fire in the compartment of fire origin and the outflow parameters that govern the smoke spread to the other locations in a building were discussed in Chapter 7. The outflow parameters include the *exhaust flow rate, temperature*, CO, CO_2 and *soot concentrations*. These outflow parameters can be used as input to computer smoke spread models which calculate, based on fluid dynamics and heat transfer, the transport of these smoke spread parameters to the other locations in a building.

The values of the smoke spread parameters of *temperature*, CO, CO_2 and *soot concentrations* can be used to assess occupant visibility, life hazards to the occupants and the financial losses to the properties in the other locations in a building. The discussion of how the values of these smoke spread parameters can be used to assess the life risks to the occupants and the financial risks to the properties will be discussed later in Chapter 11. In this chapter, we will discuss mainly the characteristics and modelling of smoke spread, as well as the construction of the various smoke spread scenarios as a result of the success and failure of *smoke control systems*.

9.2.1 Buoyancy Driven Flow

Smoke spread is driven by the *buoyancy force* which is present because smoke is lighter than air. The hotter is the smoke, the lighter is the smoke and the stronger is the buoyancy force that drives it upward. In a building fire situation, smoke typically spreads along the ceiling of any open horizontal spaces first, such as the compartments and corridors, and then seeks vertical conduits, such as the stairwells and elevator shafts, to rise to the upper floors of a building. If the fire is in an atrium, smoke rises first to fill the upper part of the atrium and then spreads horizontally through openings to the other floors. The buoyancy effect is illustrated in Figure 9.1 where the outflow of smoke hugs the upper part of an opening in a compartment fire.

The buoyancy force is present because the density of smoke is lighter than that of the ambient air, as described by Equation 9.1.

$$\frac{\partial p}{\partial z} = (\rho_g - \rho_a)\, g. \tag{9.1}$$

In Equation 9.1, $\partial p / \partial z$ is the buoyancy force per unit volume of smoke ($N \cdot m^{-3}$) in the upward direction (z), ρ_g is the smoke density ($kg \cdot m^{-3}$), ρ_a is the air density ($kg \cdot m^{-3}$) and g is the gravitational acceleration ($m \cdot s^{-2}$). Equation 9.1 shows that the larger the density difference is, the larger the buoyancy force is.

Table 9.1 shows the density values of air at various temperatures, from 300 to 1400 K (27–1127 °C). The temperature range covers the

Figure 9.1 Outflow of smoke through the upper part of an opening in a compartment fire.

Table 9.1 Thermophysical properties for air at standard
atmospheric pressure (source: SFPE Handbook, SFPE, 2002,
Table B.2).

Temperature (K)	P (kg·m^{-3})	C_p (kJ·kg^{-1}·K^{-1})
300	1.1774	1.0057
400	0.8826	1.0140
500	0.7048	1.0295
600	0.5879	1.0551
700	0.5030	1.0752
800	0.4405	1.0978
900	0.3925	1.1212
1000	0.3524	1.1417
1100	0.3204	1.1600
1200	0.2947	1.1790
1300	0.2707	1.1970
1400	0.2515	1.2140

In Table 9.1, ρ is the density and C_p is the specific heat at constant
pressure.

typical range of temperatures in a compartment fire (see Figure 8.2 in
Chapter 8). If the smoke density can be approximated by the air density,
Table 9.1 can be used to calculate the smoke densities at difference
temperatures. Table 9.1 shows that the smoke density at the high fire
temperature of 1400 K drops to a small fraction of its value at the
ambient temperature of 300 K. At 1400 K, the smoke density drops
to 21 % of its value at ambient temperature. At this temperature, the
buoyancy force on a unit volume of smoke is 4.7 times its weight. The
buoyancy force is the dominant driving force in smoke spread. There
are also other driving forces, such as stack effect and wind effect in a
building, which will be discussed in the next two sections.

9.2.2 Stack Effect

Smoke spread can be affected by natural convective flow that may be
present in a building. Natural convective flow is present in a building
when the indoor temperature is different from the outdoor temperature.
Figure 9.2 shows the indoor and outdoor pressure profiles when the
indoor temperature is higher than the outdoor temperature. Outdoor
air flows into the building through openings on the exterior walls in
the lower floors, rises to the upper floors through vertical shafts and

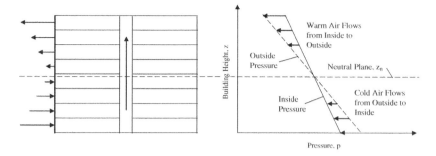

Figure 9.2 Natural convection in a building as a result of the temperature difference between the indoor and the outdoor.

exits the building through openings on the exterior walls in the upper floors.

The indoor pressure profile adjusts naturally its position relative to the outdoor profile so that the outflow from the upper floors balances the inflow from the lower floors. If the openings in the upper floors are the same as those in the lower floors, the neutral plane is close to the mid height of the building (not exactly at the mid height because the density of the air going into the building and the density of the air coming out of the building are not exactly the same due to differences between indoor and outdoor temperatures). If there are more openings in the upper floors, the neutral plane is above the mid height of the building. If the opposite is true, the neutral plane is below the mid height of the building.

In this case when the indoor temperature is higher than the outdoor temperature, as is normally the case in the winter time when the indoor is heated, the flow upward of warm air in the vertical shafts is commonly referred to as the *stack effect*. If the indoor temperature is lower than the outdoor temperature, as is normally the case in the summer time when air conditioning is on, the flow reverses and moves downward in the vertical shafts. The flow downward of cool air in the vertical shafts is referred to as the *reverse stack effect*. The stack effect and the reverse stack effect in a building can enhance or impede smoke movement upward in vertical shafts.

The driving force of natural convection is the pressure difference between the indoor and outdoor. Using Figure 9.2, the equation for the pressure difference is

$$(p_i - p_0) = g\,(\rho_0 - \rho_i)\,(z - z_n). \tag{9.2}$$

In Equation 9.2, p_i is the indoor pressure (N·m^{-2}), p_o is the outdoor pressure (N·m^{-2}), g is the gravitational acceleration (m·s^{-2}), ρ_o is the outdoor air density (kg·m^{-3}), ρ_i is the indoor air density (kg·m^{-3}), z is the vertical coordinate (m) and z_n is the height of the neutral plane (m). The densities in Equation 9.2 can be replaced with temperatures by utilizing the ideal gas law.

$$p = \rho \, R_a \, T \tag{9.3}$$

In Equation 9.3, p is the pressure (N·m^{-2}), ρ is the density (kg·m^{-3}), R_a is the specific gas constant for air (N·m·kg^{-1}·K^{-1}) and T is the temperature (K). Utilizing Equation 9.3, Equation 9.2 can be rearranged into:

$$(p_i - p_0) = \frac{g \, p_i}{R_a} \left(\frac{1}{T_0} - \frac{1}{T_i} \right) (z - z_n). \tag{9.4}$$

In Equation 9.4, T_o is the outdoor temperature (K) and T_i is the indoor temperature (K). The indoor air pressure p_i is very close, in reality, to the value of the standard pressure at sea level p_s, with only a small difference that varies with height. The indoor pressure p_i on the right hand side of the equation can be replaced with p_s, which gives:

$$(p_i - p_0) = \frac{g \, p_s}{R_a} \left(\frac{1}{T_0} - \frac{1}{T_i} \right) (z - z_n). \tag{9.5}$$

With the gravitational acceleration $g = 9.807$ m · s^{-2}, the standard pressure at sea level $p_s = 101.3 \times 10^3$ N·m^{-2} and the specific gas constant for air $R_a = 287.1$ N·m·kg^{-1}·K^{-1}, Equation 9.5 becomes:

$$(p_i - p_0) = 3460 \left(\frac{1}{T_0} - \frac{1}{T_i} \right) (z - z_n). \tag{9.6}$$

In Equation 9.6, p_i and p_o have units of Newton per square meter, T_o and T_i have units of Kelvin and z and z_n have units of meter. Equation 9.6 shows that for typical indoor and outdoor temperatures, the pressure difference is only significant when the building is very tall. For example, assuming an indoor temperature of 20 °C and an outdoor temperature of −20 °C, a six-storey low-rise building with a height of 20 m has a maximum pressure difference of 19 N·m^{-2}. A 60-storey high-rise building with a height of 200 m and the same indoor and outdoor temperatures, on the other hand, has a more significant maximum pressure difference of 187 N·m^{-2}.

9.2.3 Wind Effect

Smoke spread can also be affected by *wind effect*. Figure 9.3 shows the effect of wind on a building, causing positive pressure on the windward side and negative pressure on the leeward side. The wind enters the building on the windward side and exits the building on the leeward side. The mainly horizontal movement of the wind can affect the smoke movement in the building. If the compartment of fire origin is on the windward side, the wind effect would enhance the smoke spread in the building. If the compartment of fire origin in on the leeward side, the wind effect would impede the smoke spread coming out from the compartment.

The wind pressure on the wall of a building is proportional to the dynamic pressure of the wind, as expressed by Equation 9.7.

$$(p_w - p_l) = (C_w - C_l) \left(\tfrac{1}{2}\rho u^2\right) \tag{9.7}$$

In Equation 9.7, p_w is the wind pressure on the windward side of the wall (N·m^{-2}), p_l is the wind pressure on the leeward side of the wall (N·m^{-2}), ρ is the air density (kg·m^{-3}) and u is the wind velocity (m·s^{-1}). Also in Equation 9.7, C_w is a positive coefficient that relates to the conversion of the wind dynamic pressure to static pressure on the windward side; and C_l is a negative coefficient that relates to the recovery of the dynamic pressure to static pressure on the leeward side. The values of C_w and C_l depend on wind directions relative to the building as well as other buildings in the vicinity. The windward coefficient C_w has a value up to

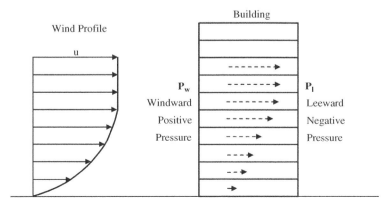

Figure 9.3 Wind effect causing positive pressure on the windward side and negative pressure on leeward side.

0.8 and the leeward coefficient C_l has a value down to -0.8 (Tamura, 1994; Klote, 2002).

To give a perspective on how much the wind effect is, assume a moderate wind velocity of $20\,\text{km}\cdot\text{h}^{-1}$, an air density of $1.18\,\text{kg}\cdot\text{m}^{-3}$ and both C_w and C_l have their maximum values of 0.8 and -0.8, respectively. Equation 9.7 shows the pressure difference across the building $(p_w - p_l)$ is $29\,\text{N}\cdot\text{m}^{-2}$. The wind effect is proportional to the square of the wind velocity. If the wind velocity is doubled to $40\,\text{km}\cdot\text{h}^{-1}$, the pressure across the building is quadrupled to $116\,\text{N}\cdot\text{m}^{-2}$. These values are comparable to those due to the stack effect (see previous section).

9.2.4 Smoke Spread Rate

The rate of smoke spread in a building depends on the driving forces, such as the buoyancy force, and flow restrictions, such as whether the doors to the corridors and stairwells are open or not. The amount of smoke that is spread, on the other hand, depends on the amount of smoke that is generated. The amount of smoke that is generated in a compartment fire can be substantial, as is often observed in building fires and fire experiments (see Figure 2.3 in Chapter 2 and Figure 9.1). In this section, we will discuss the amount of smoke that is generated in a typical compartment fire, which gives us a perspective of the amount of smoke that is involved in smoke spread.

The amount of smoke that is generated in a compartment fire is related to the amount of heat that flows through its openings, such as doors and windows. In a compartment fire, the heat that is released is dissipated through: (1) conduction through the boundary walls, (2) radiation through openings and (3) convection through openings. The convective component of the heat dissipation governs the volume of smoke that flows through the openings. The relationship is described by Equation 9.8.

$$V = \frac{Q_c}{\rho_g\,C_p\,(T_g - T_a)} \tag{9.8}$$

In Equation 9.8, V is the volume of smoke that flows through the openings (m^3), Q_c is the convective heat loss through the openings (kJ), ρ_g is the smoke density $(\text{kg}\cdot\text{m}^{-3})$, C_p is the specific heat of the smoke at constant pressure $(\text{kJ}\cdot\text{kg}^{-1}\cdot\text{K}^{-1})$, T_g is the smoke temperature (K) and T_a is the ambient temperature (K).

The volume of smoke generated can be calculated using the above Equation 9.8 and the thermophysical properties in Table 9.1. Assuming, for this discussion, a convective heat loss Q_c of 3×10^5 kJ, which can be the result of a convective heat loss rate of 1 MW for 5 minutes. The smoke temperature T_g exiting from a compartment fire is typically in the range of 1000–1300 K. If we assume an ambient temperature of 300 K, the calculated smoke volume is in the range of 926–1065 m^3. This is a large volume of smoke. Consider the typical cross-sectional area of a corridor is about 6 m^2 and a stairwell about 8 m^2, the hot smoke generated can fill a 166 m long corridor or a 124 m tall stairwell. The large amount of smoke generated pose significant hazards to the occupants and property. It is, therefore, important to have smoke control systems in place to prevent such smoke spread. This will be discussed in later sections of this chapter. Risk to life and property depend on how reliable and effective these smoke control systems are.

9.2.5 Smoke Spread Models

We have discussed in the previous sections the major forces that drive the smoke spread, and the large amount of smoke that is typically involved. The actual calculation of smoke spread in a building is usually conducted with the help of computer smoke spread models. In this section, we will discuss some of these models. Computer smoke spread models use fluid dynamics and heat transfer to calculate the transport of the time-dependent values of the previously mentioned smoke spread parameters of *temperature*, CO, CO_2 and *soot concentrations* to every location in a building.

Computer smoke spread models include both zone and field models. The zone models divide the building space into multiple zones and calculate the averaged values of the smoke spread parameters in each zone. A zone can be a compartment, a corridor or a stairwell. The field models, on the other hand, divide the whole building into small cells and calculate the averaged values in each cell. The cells can be very small and a field model therefore can calculate the values of the smoke spread parameters at every point in a building.

Whether they are zone models or field models, these computer smoke spread models calculate the transport of the time-dependent values of the smoke spread parameters of *temperature*, CO, CO_2 and *soot concentrations* to every location in a building. These values will be used, as will be discussed in later chapters, for the assessments of occupant visibility (based on soot concentration), life hazards to the occupants (based on temperature and CO and CO_2 concentrations), as well as financial losses to properties (based on smoke and heat damages). Together with the probabilities of smoke spread scenarios, which will be discussed in the next section, these smoke spread values can be used to assess the life risks to the occupants and financial risks to the properties.

During the past 20 years, many computer smoke spread models have been developed by various organizations, which include both zone and field models (Olenick and Carpenter, 2003). One of the well-known zone models is the *CFAST* which was developed at the National Institute of Standards and Technology (NIST) (Peacock *et al.*, 2005). Another zone model of interest is the *NRCC Smoke Spread Model* which was developed at the National Research Council Canada (NRCC) (Hadjisophocleous and Yung, 1992; Hokugo, Yung and Hadjisophocleous, 1994). The NRCC model is a simple smoke spread model specifically designed with a fast computational time to allow for comprehensive, multi-scenario, fire risk assessments.

With the advancement of faster and more powerful computers, field models based on computational fluid dynamics (CFD) are becoming more popular. These models are especially suitable for applications in large spaces, such as atrium or auditorium, where zone models are not appropriate. Each of these field models was developed with certain applications in mind. Some are for general applications and some are developed specifically for fire applications. Those for fire applications use turbulence models that are more suitable for buoyancy driven flows.

In this section, we will mention briefly two well-known CFD models. The first is the *PHOENICS* model (parabolic hyperbolic or elliptic numerical integration code series) which has been developed for over 25 years by CHAM in the UK (PHOENICS, 2007). PHOENICS is a general purpose code. Figure 9.4 shows the output of the PHOENICS in a study by Chen and Yung (2007) to model the design of smoke extraction strategies for a new fire laboratory in Australia. They used the model to study the smoke build-up in the burn hall and whether the external cross winds would force the smoke to leak out through the fresh air intake louvers around the lower perimeter of the laboratory.

Another popular CFD model is the *fire dynamic simulator* (*FDS*), which was also mentioned in Chapter 7. The FDS, specifically developed by NIST for fire and smoke simulations, employs the large eddy simulation (LES) technique to model the turbulence in buoyancy driven flows suitable for fires (McGrattan and Forney, 2006). Figure 9.5 is an

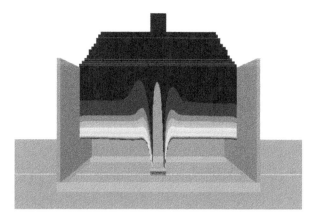

Figure 9.4 PHOENICS output of the smoke build-up in a fire laboratory with a 5 MW fire in the centre, a smoke extraction rate of $39 \, m^3 \, s^{-1}$ at the top and an external cross wind of $6 \, m \, s^{-1}$ from left to right (from Chen and Yung, 2007, reproduced by permission of the Society of Fire Protection Engineers).

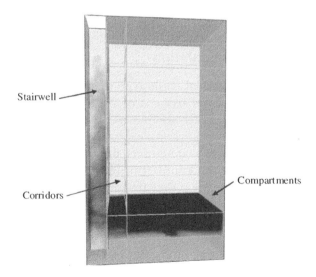

Figure 9.5 FDS output of the smoke spread from a compartment fire to the corridor and then up the stairwell (figure courtesy and by permission of Dr Yunlong Liu of Sydney, Australia).

illustration of the output of FDS for the smoke flow in a building with corridors and stairwells.

It should be mentioned here that fire modelling is not the major focus of this book; whereas the use of these models in conjunction with the smoke spread scenarios is. For more details on fire models, readers should consult the references.

9.3 Smoke Control Systems to Clear Smoke in Evacuation Routes

Smoke spread from the compartment of fire origin to other locations in a building poses life risks to the occupants and financial risks to properties in these other locations. To minimize the risks to the occupants and properties, smoke control systems are normally required in building regulations to prevent smoke spread. There are basically three types of smoke control strategies. They are described in the following sections.

9.3.1 Door Self-closers and Automatic Shut-offs of Mechanical Ventilation

The first type of smoke control is the type that is employed to prevent smoke from leaving the compartment of fire origin and also prevent smoke from entering into other compartments or stairwells. This type of smoke control includes *door self-closers* and automatic shut-offs of mechanical ventilation. Door self-closers prevent smoke from coming out through the door of the compartment of fire origin and prevent smoke from entering into other compartment or stairwells. The success of this smoke control device depends on the proper design of the force that is exerted by the self-closer. If the force is not strong enough, it may not close the door properly. If the force is too strong, it may be difficult for the occupants to open the door. Automatic shut-offs of mechanical ventilation prevents smoke spread through the ventilation system, if such ventilation system is used to provide ventilation to individual compartments. The success of this system depends on proper design and maintenance of smoke detection in the ventilation system, automatic shut-off of the fans and activation of the dampers. The probabilities of success and failure of door self-closers and automatic shut-offs of mechanical ventilation can be used to create smoke spread scenarios, which will be discussed in Section 9.3.4.

9.3.2 Automatic Fire Floor Smoke Extraction and Pressurization of Floors Above and Below

The second type of smoke control strategy is the type that is employed to exhaust the smoke from the fire floor and to prevent the smoke from migrating to the floor above and below. This clears the smoke on the fire floor and also the floors above and below for occupant evacuation. This system includes exhaustion of smoke from the fire floor to the outside and to pressurize the floors above and below, as depicted in Figure 9.6. The probability of success depends on proper design and maintenance of the smoke detection and mechanical exhaustion and pressurization. The probabilities of success and failure of this automatic *smoke extraction* and pressurization system can be used to create smoke spread scenarios, which will be discussed in Section 9.3.4.

9.3.3 Automatic Stairwell and Elevator Shaft Pressurization

The third type of smoke control strategy is the type that is employed to prevent the smoke from entering the stairwells and elevator shafts. This allows safe evacuation in the stairwells and elevators and prevents smoke spread through stairwells and elevator shafts. (The use of 'safe elevators' for evacuation is an option being considered by the fire community.) This system uses fans to pressurize the stairwells and elevator shafts, as depicted in Figure 9.7. The success of this smoke control device depends on the proper design of the fans to provide a uniform and proper level of pressurization in the stairwell. If the pressure is not high enough,

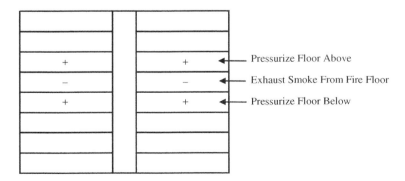

Figure 9.6 The second type of smoke control strategy is to exhaust the fire floor and to pressurize the floors above and below.

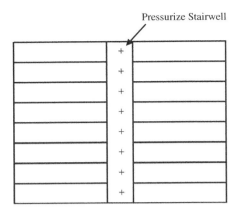

Figure 9.7   The third type of smoke control strategy is to pressurize the stairwell.

it may not stop the smoke from entering the stairwell. If the pressure is too high, it may be difficult for the occupants to open the doors. Figure 9.8 is a photo of the unique ten-storey facility at the National Research Council Canada that was built for smoke spread and smoke control studies. Much experimental work on smoke control was conducted using this facility (Tamura, 1994; Klote, 2002). The probabilities of success and failure of this automatic *stairwell and elevator shaft pressurization* system can be used to create smoke spread scenarios, which will be discussed in Section 9.3.4.

9.3.4 Smoke Spread Scenarios

The probability of success or failure of the above three types of smoke control systems can be used to create a set of *smoke spread scenarios* for fire risk assessments. These three smoke control systems are: (1) door self-closers and automatic shut-offs of mechanical ventilation, (2) automatic fire floor smoke extraction and pressurization of the floors above and below and (3) stairwell and elevator shaft pressurization. Figure 9.9 shows a total of eight possible smoke spread scenarios as a result of the various combinations of these three smoke control systems. Two of these eight scenarios allow smoke spread; the other six do not. Scenario G allows smoke spread to the corridor on the fire floor only, whereas Scenario H allows smoke spread to both the corridor and the stairwells and elevator shafts. It should be noted that in these scenarios, the extraction and pressurization system is assumed to be able to prevent

Figure 9.8 The ten-storey experimental tower at the National Research Council Canada for smoke spread and smoke control studies (photo by author, reproduced by permission of the National Research Council Canada).

smoke spread to other locations even when the mechanical ventilation is not shut off (no recirculation with fresh air from outside and all exhaust air goes outside).

In Figure 9.9, the probabilities of success and failure of these smoke control systems depend very much on the proper design and maintenance of these systems. Different from other fire protection systems, such as alarms or sprinklers, statistical values of the reliability and effectiveness of these systems are not easily available. In the absence of available data, fire safety engineers and authorities having jurisdiction should agree on the values to be used in fire risk assessment. The values should be supported by engineering design and analysis to show that these systems can work effectively, commissioning tests to confirm that they work as they are supposed to, and an adequate maintenance schedule to ensure that they work reliably. Reliability and

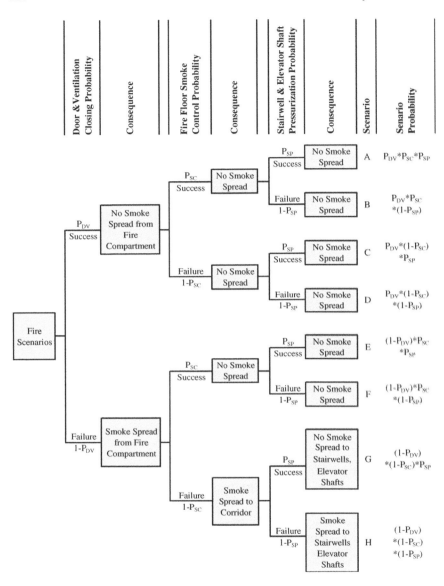

Figure 9.9 Smoke spread scenarios based on probabilities of success or failure of smoke control systems. *Note*: P_{DV} = probability of success of closing doors and shutting off ventilation systems, P_{SC} = probability of success of extracting smoke on the fire floor and pressurizing the floors above and below, P_{SP} = probability of success of pressurizing stairwells and elevator shafts.

effectiveness of fire protection systems will be discussed in more detail in Chapter 13.

The probabilities of the smoke spread scenarios, together with the time-dependent values of the smoke spread parameters of *temperature*, CO, CO_2 and *soot concentration*, will be used to assess the life risks to the occupants and financial risks to the properties. This will be discussed in Chapter 11.

9.4 Summary

The development of a fire in a compartment poses not only physical harm to the occupants and properties in the compartment of fire origin, but also risks to the occupants and properties in the other locations in a building. The outflow of heat, toxic gases and smoke from the compartment of fire origin can spread quickly to the other locations in a building, posing life risks to the occupants and financial risks to the properties in these other locations. The spread of heat, toxic gases and smoke in a building is commonly referred to as *smoke spread*. Smoke spread, therefore, refers to the spread of not only smoke particles but also toxic gases and heat.

Smoke spread in a building is governed by buoyancy force, stack effect and wind effect. The actual calculation of the smoke spread is usually conducted using computer smoke spread models. To minimize the risks to the occupants and properties, smoke control systems are normally required in building regulations to prevent smoke spread. There are basically three types of smoke control strategies: (1) door self-closers and automatic shut-offs of mechanical ventilation, (2) automatic fire floor smoke extraction and pressurization of the floors above and below and (3) stairwell and elevator shaft pressurization. Based on the success or failure of these control systems, smoke spread scenarios can be constructed. The probabilities of the smoke spread scenarios, together with the time-dependent values of the smoke spread parameters of *temperature*, CO, CO_2 and *soot concentration*, can be used to assess the life risks to the occupants and financial risks to the properties.

The probabilities of success and failure of these smoke control systems depend very much on the proper design and maintenance of these systems. Reliability and effectiveness of fire protection systems will be discussed in more detail in Chapter 13.

9.5 Review Questions

9.5.1 Calculate the probabilities of all the smoke spread scenarios in Figure 9.9. Assume the probability of success of door and ventilation closing $P_{DV} = 0.5$, the probability of fire floor smoke control $P_{SC} = 0.5$ and the probability of stairwell and elevator shaft pressurization $P_{SP} = 0.5$.

9.5.2 Calculate the probability of the scenario of smoke spread to all locations (Scenario H) in Figure 9.9. Assume higher probabilities of success for door and ventilation closing, $P_{DV} = 0.9$, for fire floor smoke control, $P_{SC} = 0.9$ and for stairwell and elevator shaft pressurization, $P_{SP} = 0.9$. Compare this probability value with the one for the same scenario in Question 9.5.1.

References

Chen, Z.D. and Yung, D. (2007) Numerical study of two air intake strategies for a new fire laboratory. *Journal of Fire Protection Engineering*, **17**(1), 27–40.

Hadjisophocleous, G.V. and Yung, D. (1992) A model for calculating the probabilities of smoke hazard from fires in multi-storey buildings. *Journal of Fire Protection Engineering*, **4**(2), 67–80.

Hokugo, A., Yung, D. and Hadjisophocleous, G.V. (1994) Experiments to validate the NRCC smoke movement model for fire risk-cost assessment, *Proceedings of the Fourth International Symposium on Fire Safety Science*, Ottawa, Canada, June 1994, pp. 805–16.

Klote, J.H. (2002) Smoke control, in *SFPE Handbook of Fire Protection Engineering 2002*, 3rd edn, Section 4, Chapter 12, National Fire Protection Association, Quincy, MA.

McGrattan, K.B. and Forney, G. (2006) *Fire Dynamics Simulator (Version 4): user's Guide*, NIST Special Publication 1019, National Institute of Standards and Technology, Gaithersburg, MD.

Olenick, S.M. and Carpenter, D.J. (2003) An updated international survey of computer models for fire and smoke. *Journal of Fire Protection Engineering*, **13**(2), 87–110.

Peacock, R.D.,Jones, W.W., Reneke, P.A. and Forney, G.P. (2005) *CFAST: Consolidated Model of Fire Growth and Smoke Transport (Version 6): user's Guide*, NIST Special Publication 1041, National Institute of Standards and Technology, August 2005, Gaithersburg, MD.

PHOENICS on-line information (2007) Concentration, Heat and Momentum Limited (CHAM), Wimbledon, U.K., http://www.cham.co.uk/website/new/phoenic2.htm.

SFPE (2002) Thermophysical property values for gases at standard atmospheric pressure, in *SFPE Handbook of Fire Protection Engineering 2002*, 3rd edn, Table B.2, National Fire Protection Association, Quincy, MA, pp. A23–A24.

Tamura, G.T. (1994) *Smoke Movement and Control in Highrise Buildings*, National Fire Protection Association, Quincy, MA.

10

Occupant Evacuation Scenarios

10.1 Overview

A fire's development in a compartment not only poses physical harm to the occupants and properties in the compartment of fire origin, but also risks to occupants and properties in other parts of the building. Heat, toxic gases and smoke from the compartment of fire origin can spread quickly to the other locations in a building. They can pose life risks to occupants and financial risks to properties in these other locations. Therefore, the occupants' safety depends on their timely evacuation to a place of safety, whether it be a refuge area inside the building or an open space outside the building, before the arrival of the critical smoke conditions in the evacuation routes that prevent evacuation. The aim of timely evacuation is to minimize the required evacuation time to ensure that it is less than the available evacuation time.

The characteristics of occupant evacuation and modelling are discussed in this chapter. Characteristic times include detection and warning time, delay start time and movement time. Also discussed are fire protection measures that can help minimize the required evacuation time by providing early fire detection and warning as well as expedite occupant response and evacuation. These safety measures include smoke alarms, live voice communication, and occupant evacuation planning, training and drills. Occupant evacuation scenarios can be constructed according to whether these fire protection measures succeed or fail. The life risks to occupants can be assessed by utilizing the probabilities of the occupant

Principles of Fire Risk Assessment in Buildings   D. Yung
© 2008 John Wiley & Sons, Ltd

evacuation scenarios and time-dependent calculation of the occupants' movements.

10.2 Occupant Evacuation Characteristics and Modelling

As was described in previous chapters, the development of a fire in a compartment poses not only physical harms to the occupants and properties in the compartment of fire origin, but also potential fire risks to the occupants and properties in the other locations in a building. The outflow of heat, toxic gases and smoke from the compartment of fire origin to the other locations in a building, commonly referred to as smoke spread, poses life-loss risks to the occupants and financial-loss risks to the properties in these other locations. Safety of the occupants, therefore, depends on timely evacuation of the occupants to a safe place, whether an open space outside the building or a refuge area inside the building, prior to the arrival of the *critical smoke conditions* in the evacuation routes that prevent evacuation. Any occupants who cannot evacuate in time and are trapped in certain locations in the building are at risk of losing their lives unless the fire department can respond and rescue them in time.

The characteristics and speed of smoke spread under various smoke spread scenarios, as a result of the success and failure of smoke control systems, were discussed in the previous chapter, Chapter 9. In this chapter, we will discuss the characteristics and speed of occupant evacuation under various occupant evacuation scenarios, as a result of the success and failure of the fire protection measures to help expedite evacuation. Those occupants who cannot evacuate before the arrival of the critical smoke conditions in the escape routes that prevent evacuation are trapped in their locations and are at risk of losing their lives unless the fire department can respond and rescue them in time. The assessment of the life-loss risks to the trapped occupants will be discussed in the next chapter, Chapter 11, based on the length of their exposure to untenable conditions before the arrival of the fire department.

The objective of a timely evacuation is to evacuate the occupants before the arrival of the *critical smoke conditions* in the evacuation routes that prevent evacuation. If the occupants are trapped in their locations, they face the impending arrival of the untenable conditions that can cost their lives, unless the fire department can respond and rescue them in time. The time duration from ignition in the compartment of fire origin to the arrival of the *critical smoke conditions* in the evacuation routes that prevent evacuation is called the *available evacuation time*, also called the *available safe egress time* (ASET). The time duration from ignition in the compartment of fire origin to the time required for the evacuation of all of the occupants is called the *required evacuation time*, also called the *required safe egress time* (RSET). The objective of a timely evacuation is to have the *required evacuation time* less than the *available evacuation time*. The essence of this evacuation objective is expressed in the following equation:

Required evacuation time < available evacuation time.   (10.1)

The *critical smoke conditions* in the evacuation routes that prevent evacuation are usually related to the arrival of a head-level smoke layer, a high level of heat and a low level of visibility, which deter occupants from attempting evacuation. There are no universally accepted standard values for these *critical smoke conditions*. Table 10.1 shows one example of these *critical smoke conditions* which are recommended by the British Standards (BSI, 2002, Part 0). The British Standards recommend that the arrival of any one of these three *critical smoke conditions* in Table 10.1 can prevent evacuation. More severe untenable conditions that can cause deaths in a short time, which include the toxic gas concentrations in addition to the heat and smoke, will be discussed in the next chapter, Chapter 11. In Chapter 11, the life-loss risks to the trapped occupants will be assessed based on the length of their exposure to these untenable conditions prior to the arrival of the fire department.

In Table 10.1, the smoke layer height of 2 m relates to a height that is just above the head of a typical person, posing an immediate threat. The temperature of 200 °C in the smoke layer relates to a threshold temperature above which the radiant heat from it ($2.5\,\text{kW}\,\text{m}^{-2}$) can cause a severe skin pain to a person (SFPE HB, 2002, Table 2–6.18). The visibility of 10 m relates to a threshold distance less than which some of the people would turn back after moving through smoke. Research on evacuation through smoke has shown that the shorter the visibility is, the more likely a person would turn back after moving through smoke.

Table 10.1 Critical smoke conditions that prevent evacuation (source: BSI, 2002, Part 0).

Critical smoke conditions that prevent evacuation	
Smoke layer height	<2 m
Smoke layer temperature	>200 °C
Visibility	<10 m

At a visibility of 10 m, British and United States studies have shown that 3–6 % of the occupants would start to turn back after moving through smoke (SFPE HB, 2002, Table 3–12.20).

The *available evacuation time* is the time duration from ignition in the compartment of fire origin to the arrival of the *critical smoke conditions* in the evacuation routes that prevent evacuation. The smoke conditions at any location in a building, at any time, and under various smoke spread scenarios, can be calculated using smoke spread models, as was discussed in Chapter 9. The smoke layer height and the smoke temperature are calculated by smoke spread models based on conservation equations of mass, momentum and energy. The level of visibility is calculated based on the smoke particle concentration and the light attenuation property, as is given by the following equation:

$$\frac{I}{I_0} = 10^{-DL}, \tag{10.2}$$

where I is the intensity of the light after transmitting through a smoke medium of length L (m), I_0 is the intensity of the incident light, and D is the *optical density* (m^{-1}). Equation 10.2 shows that the attenuation of light is a function of the product of the *optical density* and the distance. The higher the optical density or the longer the distance, the larger is the attenuation. At a distance that is equal to the inverse of the optical density, that is, $L = 1/D$, the light attenuation is 90 %. This distance that is equal to the inverse of the optical density is nominally used as the visibility limit. For example, if the optical density is $0.1\,m^{-1}$, then the visibility is 10 m. If the optical density is $0.5\,m^{-1}$, then the visibility is 2 m.

The optical density D depends on the mass density, size distribution and the optical property of the smoke particles through the following relationship:

$$D = \tfrac{1}{2.3} K_m \rho_s, \tag{10.3}$$

where K_m is the *specific extinction coefficient* $(m^2 \cdot g^{-1})$ and ρ_s is the *smoke mass density* $(g \cdot m^{-3})$. The specific extinction coefficient K_m has a value of $7.6\,m^2 \cdot g^{-1}$ for smoke produced during flaming combustion of wood and plastics and a value of $4.4\,m^2 \cdot g^{-1}$ for smoke produced during pyrolysis of these materials (SFPE HB, 2002, Sec 2–13). The smoke mass density is calculated by smoke spread models based on conservation equations of mass, momentum and energy.

The *required evacuation time* is the time duration from ignition in the compartment of fire origin to the time required for the evacuation of all of the occupants to a safe place. The required evacuation time consists

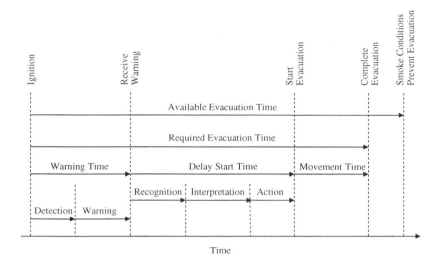

Figure 10.1 Evacuation times.

of three event times: (1) detection and warning time, (2) delay start time and (3) movement time. In addition, the delay start time also consists of three event times: (1) warning signal recognition time, (2) warning signal interpretation time and (3) pre-movement action time. All of these event times are depicted in Figure 10.1. The characteristics of all these event times and how they can be shortened by the use of fire protection measures are discussed in the following sections. The main objective of occupant evacuation planning is to try to minimize the *required evacuation time* so that it is less than the *available evacuation time*.

It should be noted that the terminology of these event times vary in the literature. The terms used here are those chosen by the author.

10.2.1 Detection and Warning Time

Fires can be detected almost immediately by the occupants if they are in the vicinity of a fire and are alert. Human senses (seeing, smelling and hearing) can detect a fire immediately if they are close to the fire and are alert. However, if they are not alert, for example, if they are asleep or intoxicated, they may not be able to detect a fire immediately even if they are in the vicinity of the fire. Although they can still be awaken by the fire and therefore discover the fire, the awakening process does not usually happen immediately. As for those occupants who are not in the vicinity of a fire, they also may not be able to detect the fire immediately. The fire cues have to travel to them first which can take some time.

Since time is of the essence in occupant evacuation, automatic detectors are normally required in building regulations to provide early detection. Detectors are typically smoke or heat detectors, which detect the smoke or heat coming out from a fire. *Smoke detectors* usually detect fires much faster than *heat detectors*. This is because smoke detectors are designed to detect the presence of a small amount of smoke particles (one reason for frequent false alarms); whereas heat detectors are designed to detect the presence of a relatively high temperature (around 70 °C to avoid false alarms from possible hot indoor conditions).

There are two types of smoke detectors: *ionization* and *photoelectric*. Ionization detectors detect smoke by sensing the change in electric current in its ionized-air chamber when smoke particles enter it and disturb the electric current. Photoelectric detectors detect smoke by sensing the change in light in its chamber when smoke particles enter it and disturb the light. Ionization detectors respond slightly faster to flaming fires; whereas photoelectric detectors respond slightly faster to smouldering fires.

Smoke detectors are designed to detect a fire at its earliest stage of development. This provides the much needed time for occupant evacuation before a fire develops into a severe fire that prevents evacuation. The time of detection can vary, depending on the growth rate of a fire (see Chapter 7). Some fires, such as flaming fires, can develop quickly from ignition to a small fire. Other fires, such as smouldering fires, can take a long time to develop from ignition to a small fire (see Table 10.2). The time of detection, therefore, is not as critical as detecting a fire when it is still small. Proper detection design includes the consideration of the type of fire expected, the type and placement of the detectors. For more details on detection design, see reference (SFPE HB, 2002, Sec 4).

Recent experimental studies by the National Institute of Science and Technology (NIST) in the USA on residential smoke alarms, employing

Table 10.2   Activation times of residential smoke alarms in the room of fire origin (source: Bukowski *et al.*, 2007).

	Activation time (min)	
	Ionization alarm	Photoelectric alarm
Flaming fires	0.5–1.2	1.3–1.8
Smouldering fires	60.3–80.4	22.3–42.5

present-day smoke alarms, residential furniture and two representative houses, have found that the activation times of smoke alarms in the room of fire origin are as shown in Table 10.2 (Bukowski *et al.*, 2007). Smoke alarms are devices that include both a smoke detector and an alarm. Table 10.2 shows that both ionization and photoelectric alarms can detect flaming fires quickly. However, the ionization alarms detect flaming fires faster than the photoelectric alarms; whereas the photoelectric alarms detect smouldering fires faster than the ionization alarms. Based on these findings, the logical conclusion to ensure early detection of all fire types is to install both types of smoke alarms.

After detection, *warning signals* are usually issued to occupants in various ways. For those occupants who are in the vicinity of a fire and are alert, warning signals are issued to them, by definition, directly and immediately. For those occupants who are in the vicinity of a fire but are not alert, warning signals are issued to them from a local alarm (if installed) directly and immediately. For those remote occupants who are not in the vicinity of a fire, warning signals are usually issued to them not directly and therefore not immediately. The warning signals are usually issued to them either through a central alarm system, or through other occupants who are aware of the fire and alert them. Table 10.3 lists the various ways warning signals are issued to occupants and the possible time delays in issuing these warning signals.

Since the main objective of occupant evacuation planning is to try to minimize the *required evacuation time* so that it is less than the

Table 10.3   Various ways, and the associated time delays, in issuing warning signals to occupants.

Warning signal	Time delays in issuing warning signals
Direct perception by occupants	No time delay
Local alarm	No time delay
Central alarm	No time delay as alarm is usually issued automatically by the activation of heat detectors
Central alarm with *voice messages*	Small time delay as alarm is usually issued after an investigation by a security staff member
Warning by others, including the use of pull stations	Time delay is unpredictable
Warning by firefighters	Time delay is based on time of arrival of firefighters

available evacuation time, every effort should be made to try to minimize the *required evacuation time*. In this case, the time required to issue warning signals after detection should be minimized as much as possible (see Figure 10.1). Table 10.3 shows that the time required to issue warning signals can be minimized if the time delays in issuing voice messages can be minimized as much as possible. One way to ensure a short time to issue warning signals is to have a proper evacuation plan and regular evacuation training sessions for the security staff so that they can issue central alarms and voice messages as quickly as possible.

The time when occupants receive their warning signals depends on where they are located. For those who are in the compartment of fire origin and are alert, they receive their warning signals almost immediately by direct perception. But for those who are located far away from the compartment of fire origin, they receive their warning signals through central alarms or warnings by others. The further the occupants are away from the compartment of fire origin, the more they are dependent on central alarms to receive early warnings.

One method to model the various times when warning signals are received by the occupants at various locations in the building is to relate the times with those of the fire development (Proulx and Hadjisophocleous, 1994). There are five characteristic states of fire development, each of which has special characteristics that can trigger certain warning signals to be issued (Hadjisophocleous and Yung, 1994). State 1 is the initial stage of fire development when the fire can be detected by one of the human senses (visual, olfactory and auditory). State 2 is the stage when sufficient smoke is generated that can trigger the activation of smoke detectors. State 3 is the stage when sufficient heat is generated that can trigger the activation of heat detectors and sprinklers. State 4 is the stage when flashover occurs which can generate significant amount of heat and smoke. State 5 is the burnout stage when the fire in the compartment of fire origin is extinguished by itself or by the firefighters.

Table 10.4 shows the occupants at every location receive various warning signals at various times. The table shows that as the fire develops, more warning signals are issued. Those who are close to the fire receive more direct warning signals early, such as direct perception and warnings from people who have seen the fire. Those who are far away receive more indirect warning signals at a later time, such as warnings from central alarms or from those who have heard from others.

Table 10.4 Warning signals received by occupants at various locations and at various states of fire growth.

State of fire growth and timelines	Warning signals received by occupants at various locations		
	Compartment of fire origin	Fire floor	Other floors
State 1: time of fire cues	Direct perception Warning by others	Warning by others	
State 2: time of Smoke detector activation	Direct perception Local alarm Warning by others	Warning by others	Warning by others
State 3: time of heat detector and sprinkler activation	Direct perception Local alarm Central alarms Voice messages Warning by others	Central alarms Voice messages Warning by others	Central alarms Voice messages Warning by others
State 4: time of Flashover	Too late	Direct perception Central alarms Voice messages Warning by others	Central alarms Voice messages Warning by others
State 5: time of Burnout	Too late	Direct perception Central alarms Voice messages Warning by others Warning by firefighters	Central alarms Voice messages Warning by others Warning by firefighters

10.2.2 Delay Start Time

After warning signals are issued to occupants, the occupants do not necessarily move immediately to a safe place. They usually do a number of things first before they decide whether to move or not. The things they do are generally categorized into three sequential events: (1) *warning signal recognition*; (2) *warning signal interpretation*; and (iii) *pre-movement action* (see Figure 10.1). The time required to do all these three things is called the *delay start time*.

Warning signal recognition is the recognition of the signal as a fire warning. Whether occupants can recognize the warning signal depends on whether occupants can hear the signal and whether they are familiar with the signal (Proulx, 2007).

One of the requirements for successful warning signal recognition is that the occupants must be able to hear the warning signal. Occupants may not hear the warning signal if the signal is not loud enough. Occupants who are asleep or intoxicated may also not hear the warning signal, especially if they are inside a dwelling unit and the warning signal is issued in a public corridor. Proper design of the warning system so that all of the occupants can hear the warning signal is critical.

The other requirement for successful warning signal recognition is that the occupants must be able to recognize the warning signal when they hear it. Occupants may not recognize the warning signal if they are not familiar with the signal. Occupants who are in their own homes or apartment units can usually recognize the warning signal immediately because they are familiar with it. Visitors to public buildings, however, may not recognize the warning signal because they are not familiar with it. To make it easier for people to recognize the fire warning signal, a distinct, universally recognizable, fire warning signal is required. In this regard, there has been some international effort to try to introduce the so-called 'temporal three' (T-3) signal as a universal fire warning signal (Proulx, 2007). It will take some time, however, before people in the world can immediately recognize this T-3 signal as a universal fire warning signal. In the meantime, evacuation training and drills for both occupants and security staff can help shorten the warning signal recognition time (see Figure 10.1).

Because of the above reasons, the time required for the occupants to recognize a warning signal is difficult to predict. It is important, however, to try to minimize the warning signal recognition time as part of the effort to try to minimize the *required evacuation time*. One important requirement to have a short warning signal recognition time is to have a properly designed warning system so that all the occupants can hear the warning signal when it is issued. Another important requirement to have a short warning signal recognition time is to have regular evacuation training sessions and drills for the occupants so that they can recognize the warning signal immediately when they hear it.

Warning signal interpretation is the interpretation of the signal into what the level of threat is and what action is required. Whether occupants can interpret the warning signal quickly as a real threat and whether they

Table 10.5   Interpretation certainty of various warning signals as fire emergencies and the associated interpretation times.

Warning signals	Interpretation certainty (that it is a fire emergency)	Interpretation time
Direct perception by occupants	High	Short
Warning by firefighters	High	Short
Warning by others	Fair	Medium
Central alarm with voice messages	Fair	Medium
Local alarm	Little	Long
Central alarm	Little	Long

have to act on it depend on what information is in the signal (Proulx, 2007). Table 10.5 shows the interpretation certainty of the various warning signals as fire emergencies, and the associated interpretation times, depends on the signal (Proulx and Hadjisophocleous, 1994). Direct perception of the fire, or warning by firefighters, have the highest certainty. If the warning signal is simply an alarm bell with no other information, occupants may not be able to interpret the signal as a real fire threat and may not know what to do. For example, they may interpret the signal as a nuisance alarm, such as a false alarm or a test alarm, and do anything. Or they may do some investigation on their own to determine whether the warning signal is real or false, which uses up some of the valuable required evacuation time. But if the warning signal includes, in addition to an alarm, a live voice message with specific information on the fire situation and specific instructions on what the occupants should do, then the occupants would interpret the signal without hesitation as a real fire threat and would follow the instructions on what to do. A live voice message is better than a pre-recorded voice message or an alarm bell because it removes any doubt that the warning signal may not be real. However, a live voice message system requires training of the security staff to provide a proper live voice message.

Because of the above reasons, the time required to interpret a warning signal is difficult to predict. It is important, however, to try to minimize the warning signal interpretation time as part of the effort to try to minimize the required evacuation time. One important requirement to have a short warning signal interpretation time is to have a properly designed warning system with a live voice message included. Another important requirement to have a short warning signal interpretation time is to have regular evacuation training sessions and drills for both the occupants and the security staff members so that the occupants can recognize the warning signal and follow the live voice instructions immediately when they hear them.

Pre-movement action is the action that occupants usually take before their movement to a safe place. Pre-movement action includes things such as putting on proper clothing, gathering of important belongings, calling the fire department, and warning others. The time required to do the pre-movement action is difficult to predict because it depends on how quickly the occupants can gather themselves to leave and whether they would take time to call the fire department and warn others. It is important, however, to try to minimize the pre-movement action time as part of the effort to try to minimize the *required evacuation time*. One important requirement to minimize the pre-movement action time is to have regular evacuation training sessions and drills for the occupants so they can plan ahead and expedite the pre-movement action.

The above discussions show the difficulties in predicting the *delay start time* accurately because of the many human factors involved. Nevertheless, some estimates of these delay start times are necessary in order to be able to do fire risk assessment. Table 10.6 shows the estimated delay start times that are suggested by the British Standards DD240 (SFPE HB, 2002, Table 3–13.1). The delay start time depends on the type of warning signal, whether occupants are awake or asleep and whether they are familiar with the building, alarm system and evacuation procedure. The table shows the importance of a properly designed warning system and the importance of regular evacuation training sessions and drills for both the occupants and security staff members.

10.2.3 Movement Time

Once an occupant has decided to move to a safe place and has done the pre-movement activities, the time required to travel to a safe place is the *occupant movement time*. The occupant movement time depends mainly on the travel speed and the travel distance. Based on the original

Table 10.6   Estimated delay start time based on the type of warning signal, whether occupants are awake or asleep and whether they are familiar with the building, alarm system and evacuation procedure (source: SFPE HB, 2002, Table 3–13.1).

Occupancy	Delay start time (min)		
	Type 1 signal	Type 2 signal	Type 3 signal
Offices, commercial and industrial buildings, schools and universities (occupants awake and familiar with the building, alarm system and evacuation procedure)	<1	3	>4
Shops, museums, leisure-sports centres, and other assembly buildings (occupants awake but may be unfamiliar with the building, alarm system and evacuation procedure)	<2	3	>6
Dormitories and residential buildings (occupants may be asleep but are predominantly familiar with the building, alarm system and evacuation procedure)	<2	4	>5
Hotels and boarding houses (occupants may be asleep and unfamiliar with the building, alarm system and evacuation procedure)	<2	4	>6
Hospitals, nursing homes and other institutions (a significant number of occupants may require assistance)	<3	5	>8

Type 1 Signal: live voice message from trained staff in control room.
Type 2 Signal: pre-recorded voice message from control room with trained staff.
Type 3 Signal: alarm signal only with non-trained staff.

works by Pauls and Fruin (SFPE HB, 2002, Equation 3), the travel speed is a function of a characteristic speed, modified by the crowd density, as is shown in the following equation:

$$S = k(1 - aD), \qquad (10.4)$$

where S is the travel speed ($m \cdot s^{-1}$), k is the characteristic speed ($m \cdot s^{-1}$), a is a constant $= 0.266$ (m^2 per person) and D is the crowd density (persons$\cdot m^{-2}$). The value of the characteristic speed k depends on a person's mobility, age, gender, and whether the evacuation route is level or a stair. For a normal adult, the k values are $1.40 \, \text{m·s}^{-1}$ for corridors and doorways, and $1.00-1.23 \, \text{m·s}^{-1}$ for stairs depending on the dimensions of the stair riser and tread (SFPE HB, 2002, Table 3-14.2). For people with mobility impairment, young children and senior adults, the k values are smaller.

Multiplying the travel speed with the crowd density gives the occupant flow per unit width of an evacuation route:

$$F = SD = k(1 - aD)D, \qquad (10.5)$$

where F is the occupant flow per unit width of the evacuation route $(\text{persons·s}^{-1}\text{·m}^{-1})$.

Using Equations 10.4 and 10.5, the travel speed and occupant flow per unit width of the evacuation route can be calculated for various evacuation routes and crowd conditions. They are shown in Table 10.7 (source: SFPE HB, 2002, Table 3–13.5). Note that these values are for normal adults with no mobility impairment. For young children, elderly, and people with mobility impairment, these values are smaller (SFPE EG, 2003, Table 6).

Table 10.7 shows that in low-rise buildings, the *occupant movement time* can be easily shorter than the *delay start time*. For example,

Table 10.7   Movement speed and occupant flow depend on evacuation route and crowd condition (source: SFPE HB, 2002, Table 3–13.5).

Escape route	Crowd condition	Occupant density (persons·m^{-2})	Movement speed[a] (m·s^{-1})	Occupant flow per width[b] $(\text{persons·s}^{-1}\text{·m}^{-1})$
Corridor	Minimum	<0.54	1.27	<0.68
Corridor	Moderate	1.08	1.02	1.09
Corridor	Optimum	2.15	0.61	1.31
Corridor	Crush	3.23	<0.30	<0.98
Stair	Minimum	<0.54	0.76	<0.41
Stair	Moderate	1.08	0.61	0.66
Stair	Optimum	2.04	0.48	0.99
Stair	Crush	3.23	<0.20	<0.66
Doorway	Moderate	1.08	0.86	0.93
Doorway	Optimum	2.37	0.61	1.44
Doorway	Crush	3.23	<0.25	<0.82

[a] Movement speed on stairs is based on the descending speed along the stair slope.
[b] Occupant flow per width of egress path is the product of the movement speed and the occupant density.

the table shows that the travel speed in stairs for a moderate crowd condition is 0.61 m s^{-1}. With that speed, the time required to descend one floor is about 10 seconds (assuming a stair slope of about 6 m and a vertical distance of about 3 m per floor). That means 2 minutes' travel in the stairs can cover 12 floors. The delay start time, as is shown in Table 10.6, can be 1–3 minutes with the best type of warning system, and 4–8 minutes with the worst type of warning system. This again shows the importance of minimizing the delay start time as much as possible in order to help make the *required evacuation time* less than the *available evacuation time*.

10.2.4 Required Evacuation Time

The required evacuation time for occupants in different locations and at different times depends on the time of detection, the time when a certain type of warning signal is received, and the corresponding *delay start time* for the type of warning signal received and, finally, the *movement time*. The time of detection is based on the five characteristic states of fire development. The time when a certain type of warning signal is received depends on how the signal is transmitted and how far the location is away from the compartment of fire origin. The *delay start time* depends on the type of warning signal that is received and the subsequent recognition, interpretation and action (RIA) process. The *movement time* depends on how far away the location is from the exit or from the refuge area.

The required evacuation time for occupants in different locations can be represented by the following time equation:

$$t[evac]_{ijk} = t[\det]_i + t[warn]_{ijk} + t[RIA]_{ijk} + t[move]_{ijk}, \qquad (10.6)$$

where $t[evac]_{ijk}$ is the required evacuation time for fire state i, warning signal type j, and occupants in location k; $t[\det]_i$ is the fire detection time at fire state i; $t[warn]_{ijk}$ is the time when warning signal type j, issued at fire state i, is received at location k; $t[RIA]_{ijk}$ is the delay start time as a result of the recognition, interpretation and action process for warning signal type j, issued at fire state i, and received at location k; $t[move]_{ijk}$ is the travel time for occupants from location k to the exit, or the refuge area, for warning signal type j, issued at fire state i, and received at location k.

Similarly, the probability whether occupants in different locations would evacuate can be represented by the following probability equation:

$$P[evac]_{ijk} = P[det]_i P[warn]_{ijk} P[RIA]_{ijk}, \qquad (10.7)$$

where $P[evac]_{ijk}$ is the probability of evacuation for fire state i, warning signal type j, and occupants in location k; $P[det]_i$ is the probability of detection at fire state i; $P[warn]_{ijk}$ is the probability of the receipt of warning signal type j, issued at fire state i, at location k; $P[RIA]_{ijk}$ is the probability of the decision to evacuate as a result of the recognition, interpretation and action process for warning signal type j, issued at fire state i, and received at location k.

Using Equations 10.6 and 10.7, one can calculate the required evacuation time for the occupants at different locations and at different times. For those occupants with a required evacuation time that is less than the available evacuation time, they can evacuate in time and are therefore safe. But for those with a required evacuation time that is longer than the available evacuation time, they can not evacuate in time and are considered trapped. These trapped occupants face the risk of life loss unless the fire department can rescue them in time.

10.2.5 Occupant Evacuation Models

We have discussed in the previous sections the various events that govern the *occupant required evacuation time*. The actual calculation of occupant evacuation in a building is usually carried out with the help of *computer evacuation models*. In this section, we will discuss some of these models. Computer evacuation models generally track the movement of individual occupants based on the attributes of individual occupants, such as mobility and speed; and the characteristics of crowd interaction, such as crowd density that affects speed or whether an occupant would queue at an exit or move on to another exit. These models primarily calculate the *movement time*. They usually required user input of detection and warning times and *delay start time*, and the required evacuation time is then calculated by adding the detection and warning time and the *delay start time* to the *movement time*.

During the past 20 years, many computer occupant evacuation models have been developed by various organizations (SFPE HB 2002, p. 3–377). One example of these models is the *Simulex model* (Thompson and Marchant, 1994). This model requires user input of detection and warning times and *delay start time*. Figure 10.2 shows a typical output of this model that predicts the movement of each occupant with time.

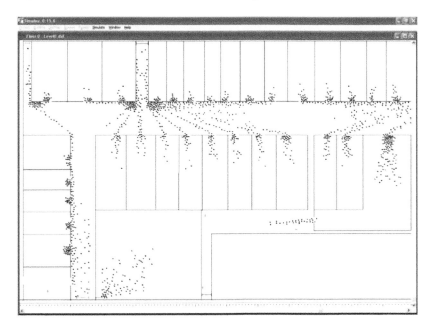

Figure 10.2   An example of the output of the computer evacuation model Simulex for a shopping centre which tracks the movement of occupants with time (each dot represents one occupant, figure courtesy and by permission of Dr Yunlong Liu of Sydney, Australia).

Another example is the National Research Council Canada's (NRCC) Occupant Response and Evacuation Models (Proulx and Hadjisophocleous, 1994; Hadjisophocleous, Proulx and Liu, 1997). The NRCC Occupant Response and Evacuation Models employ the above discussed concepts on detection and warning time, the RIA process, to model the evacuation of individual occupants at various locations in a building and at various times. These computer models were developed as sub-models of the comprehensive NRCC Risk-Cost Assessment Model (Yung, Hadjisophocleous and Proulx, 1997). Other sub-models that are part of the NRCC Risk-Cost Assessment Model include the Design Fire and Smoke Spread Models. A comprehensive risk assessment model such as the NRCC Risk-Cost Assessment Model can predict the time when the evacuation routes become untenable and consequently the number of occupants who are trapped in the building. Figure 10.3 shows the NRCC model predictions for an actual high-rise building fire that include the time of smoke detection, the time when the stairwell becomes untenable, the number of occupants who are trapped in the high-rise building when

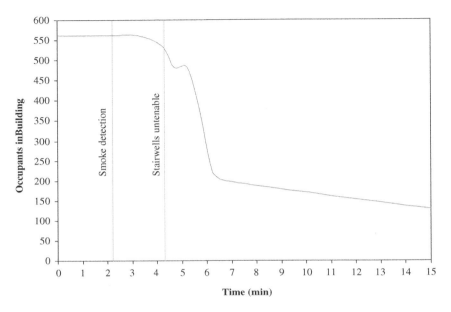

Figure 10.3 Predictions from the NRCC Risk-Cost Assessment Model for an actual highrise apartment building fire that include the time of smoke detection, the time when the stairwell becomes untenable, the number of occupants who are trapped in the building when the stairwell becomes untenable, and the evacuation profile if the stairwell does not become untenable (from Yung, Proulx and Benichou, 2001, reproduced by permission of Interscience Communications Ltd.).

the stairwell become untenable, and also the occupant evacuation profile if the stairwell does not become untenable.

It should be mentioned here that occupant evacuation modelling is not the major focus of this book; whereas the use of these models in conjunction with the occupant evacuation scenarios is. For more details on fire models, readers should consult the references.

10.3 Occupant Safety Measures to Expedite Occupant Response and Evacuation

Smoke spread from the compartment of fire origin to other locations in a building poses life risks to the occupants and monetary risks to the properties in these other locations. Safety of the occupants, therefore, depends on a timely evacuation of the occupants to a safe place, whether an open space outside the building or a refuge area inside the building, prior to the arrival of the critical smoke conditions in the evacuation routes that prevent evacuation. To provide timely evacuation of the occupants to a safe place, fire protection measures are usually required in buildings. These fire protection measures are designed to provide early fire detection and warning as well as expedite occupant response and evacuation. The objective is to minimize the *required evacuation time* so that it is less than the *available evacuation time*. These fire protection measures are described in the following sections.

10.3.1 Automatic Smoke Alarms

To provide early fire detection and warning, automatic *smoke alarms* are usually required in buildings. Smoke alarms include both fire detection and the sounding of alarms either locally or through a central alarm system. The characteristics of smoke alarms were discussed previously in Section 10.2.1. Smoke alarms, however, provide early detection and warning only if they work, that is, only if they can be activated by a fire when it is still small. If they work, they provide an early detection and warning time such as those shown in Table 10.2. If they don't, then their presence has no impact on early detection and warning.

The reliability of fire alarms is usually not a 100 %. Table 10.8 shows the United States experience with the *reliability of smoke alarm* activation in two recent five-year study periods: 1988–1992 and 2000–2004. The table shows that the reliability varies from a low of 58 % to a

Table 10.8 US experience with the reliability of smoke alarm activation against fires that should activate smoke alarms.

Occupancy	Reliability of smoke alarm activation	
	1988–1992 (Hall, 1994)	2000–2004 (Ahrens, 2007)
Dormitory and Barrack	87 %	94 %
Nursing home	86 %	92 %
Hospital	86 %	90 %
Correctional facilities	86 %	87 %
Rooming and boarding house	–	86 %
Hotel and motel	78 %	84 %
Office	–	80 %
Educational	79 %	79 %
Apartment	69 %	77 %
One and two family dwelling	68 %	67 %
Stores and mercantile	–	62 %
Public assembly	69 %	58 %

high of 94 %, depending on occupancy. The table also shows that the reliability has improved slightly from 1988–1992 to 2000–2004.

The reliability of smoke alarms has an impact on the probability of occupant evacuation scenarios with an early *detection and warning time*. Consequently, the reliability of smoke alarms has an impact on the life risks to the occupants. This will be discussed in more details in Section 10.3.5 Occupant Evacuation Scenarios.

10.3.2 Live Voice Communication

To help occupants to be able to recognize and interpret a warning signal quickly as an imminent threat, the issuance of a *live voice message* by trained security staff following the sounding of a central alarm is important. A live voice message is better than a pre-recorded voice message or an alarm bell because it removes any doubt that the warning signal may not be real. However, a live voice message works only if it provides accurate information on the fire situation and clear instructions on what the occupants should do. This requires training of the security staff to provide such a proper live voice message. The probability of success of providing such a live voice message by security staff depends on proper planning and proper training of the staff. If it works, such a live voice message would help minimize the *delay start time* to those shown as 'Type 1 Signal' in Table 10.6.

The probability of success of providing a proper live voice message depends on proper planning and proper training of the security staff. If it can be successfully exercised, then it has an impact on the probability of occupant evacuation scenarios with a short *delay start time*. Consequently, the probability of success of a proper live voice message has an impact on the life risks to the occupants. This will be discussed in more details in Section 10.3.5.

10.3.3 Evacuation Planning, Training and Drills

One way to help occupants to minimize the *delay start time* is to have regular evacuation training sessions and drills for the occupants so that they can plan ahead and expedite the pre-movement activities. Regular *evacuation training and drills* can also shorten the *movement time* to a safe place because it provides the occupants with prior knowledge of the best evacuation route to take. Better planning of evacuation routes such as better lighting in stairs and corridors will also help. Figure 10.4 shows

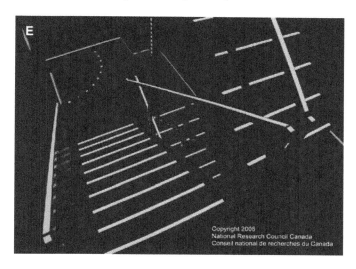

Figure 10.4 Photoluminescent material can help people see their way in evacuation routes that are without power or filled with smoke (photo courtesy of Dr Guylene Proulx, reproduced by permission of the National Research Council Canada).

the use of *photoluminescent material* can help people see their way in stairs that are without power or filled with smoke (Proulx *et al.*, 2007). It is shown here just as an example of research and development efforts worldwide to come up with better ways to help evacuate occupants in buildings in case of emergency.

Regular evacuation training and drills are especially important for high-rise buildings where controlled selective evacuation of only certain floors is used rather than the uncontrolled total evacuation of the whole building. Example of controlled selective evacuation is to evacuate all the floors above the fire floor and only one floor below the fire floor. This helps to avoid congestion in the stairs and the slow down of the evacuation process. However, controlled selective evacuation only works if the evacuation messages are clear and the occupants are willing to follow the instructions. There are scepticisms that occupants in high-rise buildings do not necessarily follow controlled selective evacuation instructions after the September 11 experience in Tower 2 of the World Trade Center. In that experience, people who didn't follow the instructions to stay in the building survived, whereas those who followed the instructions to stay perished with the collapse of the building.

Regular evacuation training and drills are also important for the security staff. This allows them to issue proper warnings and messages quickly to the occupants.

The probability of success of regular evacuation training and drills depends on whether such evacuation training and drills can be successfully implemented. If it can be successfully implemented, then it has an impact on minimizing both the *delay start time* and *movement time*. Consequently, the probability of success of regular evacuation training and drills has an impact on the life risks to the occupants. This will be discussed in more details in Section 10.3.5.

10.3.4 Refuge Areas and Safe Elevators

In a high-rise building, not all of the occupants can be easily evacuated because of the large number of people involved and the height from which they have to descend. This is especially true for those with disability. One way to help these occupants with disability is to provide *refuge areas* where they can stay temporarily until safe to leave or until they are rescued by firefighters (Proulx and Yung, 1996). Another way to help these occupants with disability is to provide *safe elevators*, which the occupants can use under the control of the firefighters to come down quickly (Kuligowski and Bukowski, 2004).

Refuge areas are areas that are designed, and more importantly perceived, to be safe from fire and smoke spread. They also need to be in areas where occupants feel comfortable that they can be easily rescued later by firefighters or other emergency responders. Areas that are often being considered are those that are part of the elevator lobby or part of the landing in the stairwell. Refuge areas are not yet common because they represent an added cost to fire protection. However, if they can be implemented, they provide added fire safety, especially to those with disability.

Safe elevators are special elevators that are designed to be safe from fire and smoke and water (from sprinkler and fire fighting operations) and have fail-safe power. These safe elevators are to be operated manually by firefighters so they can control where the elevators should stop and who can get in. The firefighters who operate the safe elevators have reliable communication with the fire command station. They are informed of the fire situation in the building and the fire command centre is informed of the situation in the elevators. Occupants waiting in the elevator lobby can also communicate with the fire command station so they are aware of the status of their impending rescue. The fire command station is also aware of the status of the occupants on each floor waiting for the rescue. Safe elevators are not yet common because they are expensive to implement. Also, there is a need to re-educate the public that it is now safe to use these special elevators for evacuation. For many years, people have been told not to use the elevators in case of fire emergency. If safe elevators can be implemented, however, they provide an added fire safety, especially to those with disability.

10.3.5 Occupant Evacuation Scenarios

The probability of success or failure of fire protection measures can be used to create a set of occupant evacuation scenarios for fire risk assessments. In this section, we will consider only the three aforementioned fire protection measures that are currently available: *automatic smoke alarm*, *live voice communication* and *evacuation planning, training and drills*. The other fire protection measures of refuge area and safe elevators are protection measures that are still being developed and may be more common in the future.

Each of these three chosen fire protection measures can help shorten the *required evacuation time*. Automatic smoke alarm can help provide early *detection and warning time* as shown in Tables 10.2 and 10.3. Live voice communication can help shorten the *delay start time* as shown in Table 10.6. Evacuation planning, training and drills can help shorten the delay start time as shown in Table 10.6, and also the *movement time* as occupants are familiar with the best routes to escape.

Figure 10.5 shows, based on the success and failure of the above three selected fire protection measures, a total of eight possible occupant evacuation scenarios. Each of these scenarios has an implied *required evacuation time*. Scenario A, with all the fire protection measures successfully operating, has the shortest required evacuation time. Scenario H, at the other extreme with all the fire protection measures fail, has the longest required evacuation time.

The probabilities of the occupant evacuation scenarios depend on the reliability and effectiveness of the fire protection measures. Reliability and effectiveness of fire protection systems will be discussed in more details in Chapter 13. The probabilities of the occupant evacuation scenarios, together with the time-dependent calculation of the movement of the occupants, can be used to assess the life risks to the occupants. This will be discussed in Chapter 11.

In Figure 10.5, the probabilities of success and failure of automatic smoke alarms depend on the design and maintenance of these systems. The probabilities of success and failure of voice communication depend

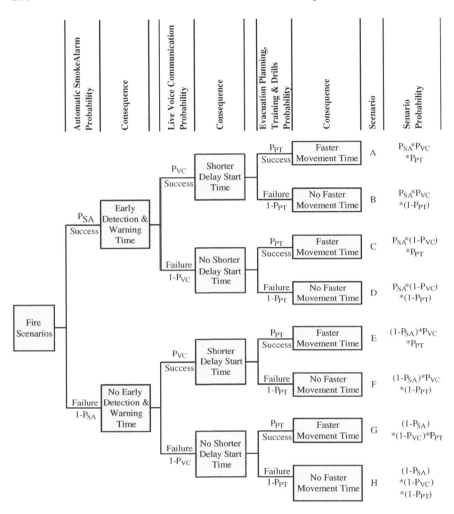

Figure 10.5 Occupant evacuation scenarios based on probabilities of success or failure of fire protection measures. *Note*: P_{SA} = probability of success of automatic smoke alarm, P_{VC} = probability of success of voice communication, P_{PT} = probability of success of evacuation planning, training and drills.

on the training of the security staff. The probabilities of success and failure of evacuation planning, training and drills depend on the preparation of an evacuation plan and the exercise of the evacuation training and drills. These probability values should be agreed upon by fire safety engineers and authorities having jurisdiction. Reliability and effectiveness of fire protection systems will be discussed in more detail in Chapter 13.

10.4 Summary

The development of a fire in a compartment poses not only physical harm to the occupants and properties in the compartment of fire origin, but also risks to the occupants and properties in the other locations in a building. The outflow of heat, toxic gases and smoke from the compartment of fire origin can spread quickly to the other locations in a building, posing life risks to the occupants and financial risks to the properties in these other locations. Safety of the occupants, therefore, depends on a timely evacuation of the occupants to a safe place, whether an open space outside the building or a refuge area inside the building, prior to the arrival of the critical smoke conditions in the evacuation routes that prevent evacuation. The objective of timely evacuation is to minimize the *required evacuation time* so that it is less than the *available evacuation time*.

Fire protection measures that can help provide timely evacuation include automatic smoke alarms, live voice communication, and evacuation planning, training and drill. These fire protection measures help minimize the *required evacuation time* by providing early fire detection and warning as well as expedite occupant response and evacuation. Other fire protection measures that are being developed include refuge areas and safe elevators. These new measures can provide added safety, especially to occupants with disability.

Based on the success or failure of these fire protection measures, occupant evacuation scenarios can be constructed. The probabilities of the occupant evacuation scenarios, together with the time-dependent calculation of the movement of the occupants, can be used to assess the life risks to the occupants.

10.5 Review Questions

10.5.1 Calculate the probabilities of all the occupant evacuation scenarios in Figure 10.5 for a flaming fire in one unit of an apartment building. The building has ionization detectors, central alarm with voice communication and regular evacuation training and drills. Assume the probability of success of voice communication is $P_{VC} = 0.9$ and the probability of success of occupant planning, training and drills is $P_{PT} = 0.5$.

10.5.2 Calculate the detection and warning time and the delay start time for all the scenarios in Question 10.5.1. Assume a detection and warning time of 5 minutes when the automatic smoke alarm is not working.

References

Ahrens, M. (2007) U.S. Experience with Smoke Alarms and Other Fire detection/Alarm Equipment, NFPA Report, April 2007, National Fire Protection Association, Quincy, MA. Tables 8–9, 26C–32C, 34C–36C.

BSI (2002) Guide to design framework and fire safety engineering procedures, *Application of Fire Safety Engineering Principles to the Design of Buildings*, PD7974-0, 2002, Part 0, British Standards Institution, London, p. 30.

Bukowski, R.W., Peacock, R.D., Averill, J.D. *et al.* (2007) *Performance of Home Smoke Alarms – Analysis of the Response of Several Available Technologies in Residential Fire Settings*, NIST Technical Note 1455, January 2007, National Institute of Standards and Technology, Gaithersburg, MD, Tables 23–24.

Hadjisophocleous, G.V., Proulx, G. and Liu, Q. (1997) Occupant Evacuation Model for Apartment and Office Buildings, National Research Council Canada Internal Report No. 741, Ottawa, May 1997, 19 pages.

Hadjisophocleous, G.V. and Yung, D. (1994) Parametric Study of the NRCC Fire Risk-Cost Assessment Model for Apartment and Office Buildings. Proceedings of the Fourth International Symposium on Fire Safety Science, June 1994, Ottawa, pp. 829–40.

Hall, J.R., Jr. (1994) The U.S. experience with smoke detectors. *National Fire Protection Association Journal*, September 1994, 36–46.

Kuligowski, E.D. and Bukowski, R.W. (2004) Design of Occupant Egress Systems for Tall Buildings. Proceedings of the CIB World Building Congress 2004, CIB HTB T3S1 Design for Fire Safety, May 2004, Toronto, pp. 1–11.

Proulx, G. (2007) Response to fire alarms. *Fire Protection Engineering Magazine*, Winter 8–14.

Proulx, G., Bénichou, N., Hum, J.K. and Restivo, K.N. (2007) Evaluation of the Effectiveness of Different Photoluminescent Stairwell Installations for the Evacuation of Office Building Occupants, Research Report 232, Institute for Research in Construction, National Research Council Canada, pp. 77, July 6, 2007, URL: http://irc.nrc-cnrc.gc.ca/pubs/rr/rr232/.

Proulx, G. and Hadjisophocleous, G.V. (1994) Occupant Response Model: A Sub-Model for the NRCC Risk-Cost Assessment Model. Proceedings of the Fourth International Symposium on Fire Safety Science, June 1994, Ottawa, pp. 841–52.

Proulx, G. and Yung, D. (1996) Evacuation Procedures for Occupants with Disabilities in Highrise Buildings. Proceedings of WOBO Fourth World Congress, November 1996, Hong Kong, pp. 1–10.

SFPE EG (2003) Human behavior in fire, *SFPE Engineering Guide*, Society of Fire Protection Engineers, Bethesda, MD, 30 pages.

SFPE HB (2002a) Behavioral response to fire and smoke, *SFPE Handbook of Fire Protection Engineering*, 3rd edn, National Fire Protection Association, Quincy, MA, Table 3-12.20, p. 3-334.

SFPE HB (2002b) Design of detection systems, *SFPE Handbook of Fire Protection Engineering*, 3rd edn, National Fire Protection Association, Quincy, MA, Sec 4, pp. 4-1–4-43.

SFPE HB (2002c) Emergency movement, *SFPE Handbook of Fire Protection Engineering*, 3rd edn, National Fire Protection Association, Quincy, MA, Equation 3 and Table 3-14.2, p. 3-370, Emergency Movement Models, p. 3-377.

SFPE HB (2002d) Movement of people: the evacuation timing, *SFPE Handbook of Fire Protection Engineering*, 3rd edn, National Fire Protection Association, Quincy, MA, Table 3-13.1, p. 3-351, Table 3-13.5, p. 3-363.

SFPE HB (2002e) Smoke production and properties, *SFPE Handbook of Fire Protection Engineering*, 3rd edn, National Fire Protection Association, Quincy, MA, Sec 2–13, p. 2-263.

SFPE HB (2002f) Toxicity assessment of combustion products, *SFPE Handbook of Fire Protection Engineering*, 3rd edn, National Fire Protection Association, Quincy, MA, Table 2-6.18, p. 2-127.

Thompson, P.A. and Marchant, E.W. (1994) Simulex: Developing New Techniques for Modelling Evacuation. Proceedings of the Fourth International Symposium on Fire Safety Science, June 1994, Ottawa, pp. 613–24.

Yung, D., Hadjisophocleous, G.V. and Proulx, G. (1997) Modelling Concepts for the Risk-Cost Assessment Model FiRECAM and its Application to a Canadian Government Office Building. Proceedings of the Fifth International Symposium on Fire Safety Science, March 1997, Melbourne, pp. 619–30.

Yung, D. Proulx, G. and Benichou, N. (2001) Comparison of Model Predictions and Actual Experience of Occupant Response and Evacuation in Two Highrise Apartment Building Fires. Proceedings of the Second International Symposium on Human Behaviour in Fire, M.I.T., Boston March 2001, pp. 77–88.

11

Fire Department Response

11.1 Overview

When people can not evacuate a building before the arrival of untenable fire and smoke conditions, they risk losing their lives if the fire department can not quickly respond and rescue them. The risk of loss of life to the trapped occupants depends on how long they are exposed to untenable fire and smoke conditions before being rescued by the fire department. In the same way, risk of property loss depends on how widely the fire and smoke may have spread in the building before the fire department are able to extinguished them.

The characteristics of fire department response are discussed in this chapter. Occupant fatalities and property loss are assessed based on the length of exposure to untenable fire and smoke conditions up to the intervention time when firefighters arrive and commence rescue and firefighting efforts. The effectiveness of firefighter's occupant rescue and fire extinguishment efforts depend on the dispatched crew size and firefighting resources, which include firefighting equipment and water resources.

Fire protection measures that can help provide early intervention time and effective occupant rescue and fire extinguishment efforts include the use of on-site security staffs and automatic notification systems to provide early notification; adequate distribution of fire stations to provide quick travel time; and adequate crew and firefighting equipment resources to provide effective occupant rescue and fire extinguishment efforts. Fire department response scenarios can be constructed on the basis of how these fire protection measures succeed or fail. The risk of life-loss to the occupants and the risk of property loss to the building can be assessed by the probabilities of the fire department response scenarios, together with the calculation of the firefighter's intervention time and their rescue and firefighting effectiveness.

11.2 Fire Department Response Time and Resources

As was described in Chapter 10, occupant safety subsequent to the development of a fire in the compartment of fire origin depends on timely evacuation of the occupants to a safe place prior to the arrival of the critical smoke conditions in the evacuation routes that prevent evacuation. A safe place can be either an open space outside the building or a refuge area inside the building that is protected from fire and smoke. Any occupants who cannot evacuate in time and are trapped in certain locations in the building face the risk of losing their lives unless the fire department can respond and rescue them in time. The risk of life-loss to the trapped occupants depends on the length of their exposure to untenable fire and smoke conditions before they are rescued by the fire department. Similarly, the risk of property loss to the building depends on the extent of fire and smoke spread in the building before they are extinguished by the fire department.

In this chapter, we will discuss the characteristics of fire department response, both time and resources, which affect occupant rescue and fire extinguishment efforts. We will also discuss how the risks of occupant life-loss and building property loss can be assessed based on the various *fire department response scenarios*. Quick response time and adequate resources help to minimize the length of exposure of any trapped occupants to untenable fire and smoke conditions before they are rescued. Quick response time and adequate resources also help to minimize the extent of fire and smoke spread in the building before they are extinguished.

11.2.1 Response Time

Subsequent to a fire ignition, a sequence of events occurs that eventually lead to the fire extinguishment and occupant rescue efforts by the firefighters at the fire scene (U.S. Fire Administration/National Fire Data Center, 2006; Benichou, Yung and Hadjisophocleous, 1999; Gaskin and Yung, 1993). This sequence of events is illustrated in Figure 11.1. They are: notification, dispatch, preparation, travel, setup and occupant rescue and fire extinguishment. Each of these events requires a certain time to carry out.

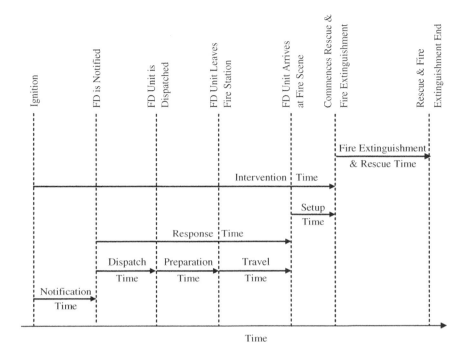

Figure 11.1   Sequential events in fire department response.

An important parameter that is often used as a measure of how quickly a fire department can respond to a fire notification is called the *response time*. The response time is defined as the duration from the time of notification of a fire to the time when the fire department arrives at the fire scene. The response time is the sum of three event times: dispatch, preparation and travel times, as shown in Figure 11.1.

Another important parameter that is used as a measure of how quickly a fire department can intervene at a fire development is called the *intervention time*. The intervention time is defined as the duration from the time of ignition of a fire to the time when the fire department commences the occupant rescue and fire extinguishment efforts. The *intervention time* includes the notification and setup times in addition to the response time, as shown in Figure 11.1.

It should be noted that the names of this sequence of events vary in the literature. The names used in this book are those that have been chosen by the author. They are also based on the names that were used in previous papers and reports co-authored by the author (Benichou, Yung and Hadjisophocleous, 1999; Gaskin and Yung, 1993).

11.2.1.1 Notification Time

Notification time is the time from ignition to the time when the fire department is notified. Notification time can vary a lot, depending to a large extent on whether the building has a security staff or an automatic notification system in place or not. If there is an on-site *security staff* who would notify the fire department immediately upon fire detector activation, or there is an *automatic notification* system in place such as a direct link of fire detector activation to the fire department, the notification time can be quite close to the fire detection time. Notification of fire detection to the security staff can be from either the activation of smoke detectors or heat detectors, whereas automatic notification to the fire department is usually from the activation of heat detectors. The activation times of smoke detectors and heat detectors were discussed in Chapter 10, with the activation time of heat detectors slightly longer than that of smoke detectors. The activation time of smoke detectors was shown in Table 10.2 to be in the range of 0.5–1.8 minutes. The quickest notification time, therefore, is in the range of 0.5–1.8 minutes. If there is no security staff or automatic notification system in place and the notification is dependent entirely on occupants who have noticed the fire and called the *fire department*, the notification time is unpredictable and can be quite long.

One way to model the various times when notifications are made to the fire department is to relate the times with those of the fire development (Proulx and Hadjisophocleous, 1994). There are five characteristic states of fire development, each of which has special characteristics that can trigger certain warning signals to be issued and subsequently certain notifications to the fire department (Hadjisophocleous and Yung, 1994). State 1 is the initial state of fire development when the fire can be detected by one of the human senses (visual, olfactory and auditory). State 2 is the state when sufficient smoke is generated that can trigger the activation of smoke detectors. State 3 is the state when sufficient heat is generated that can trigger the activation of heat detectors and sprinklers. State 4 is the state when flashover occurs which can generate significant amount of heat and smoke. State 5 is the burnout stage when the fire in the compartment of fire origin is extinguished by itself or by the firefighters.

Table 11.1 shows the various types of warning signals received by the occupants at various locations and at various states of fire development. The table also shows how the fire department may be notified at these various states of fire development. At state 1 of fire development, notification is mainly from occupants who have noticed the fire. At

Table 11.1 First notification time to the fire department based on the states of fire development.

State of fire growth and timelines	Warning signals received by occupants at various locations			First notification to fire department by
	Compartment of fire origin	Fire floor	Other floors	
State 1: time of fire cues	Direct perception Warning by others	Warning by others		Occupants
State 2: time of smoke detector activation	Direct perception Local alarms Warning by others	Warning by others	Warning by others	Occupants, security staff
State 3: time of heat detector and sprinkler activation	Direct perception Local alarms Central alarms Voice messages Warning by others	Central alarms Voice messages Warning by others	Central alarms Voice messages Warning by others	Occupants, automatic notification
State 4: time of flashover	Too late	Direct perception Central alarms Voice messages Warning by others	Central alarms Voice messages Warning by others	Occupants, passers-by
State 5: time of burnout	Too late	Direct perception Central alarms Voice messages Warning by others	Central alarms Voice messages Warning by others	Occupants, passers-by

state 2, notification can be from the occupants or the security staff if they are present. At state 3, notification can be from the occupants or the automatic notification system if they are in place. At later states of fire development, notification is mainly from the occupants or the passers-by who have noticed the fire and called the *fire department*.

The probability of notification at the early states of fire development is increased with the presence of a security staff or the installation of an automatic notification system. Without the security staff or the automatic notification system, the probability of notification to the fire department depends on the types of warning signals that are received by the occupants at various locations and the subsequent recognition, interpretation and action (RIA) process that was discussed in Chapter 10. The probability of early notification by occupants is increased with the installation of automatic detection and central alarm systems so that any fire is detected early and more people are aware of it.

The probability of notifying the fire department, at each state of fire development, can be modelled based on the probability of fire detection and the probability that some one would notify the fire department subsequent to the receipt of warning signals (Hadjisophocleous and Yung, 1994). The formulation of the probability of notification at each state of fire development is given below. The probability of notifying the fire department at each state of fire development depends on the types of warning signals that are received by the occupants at various locations and the subsequent RIA process that affects the decision whether to evacuate and whether to notify the fire department (discussed in Chapter 10). The detailed modelling of the probability of notification is very complex because of the involvement of the modelling of human behaviour of many people in an emergency. The probability of notification at each state of fire development, however, can be assumed based on the consideration of the general level of warning that has been received by the occupants and the probability that someone out of all the occupants would call.

At state 1 of fire development, the probability of notifying the fire department for the very first time, $P(firstcall)_1$, is given by the following equations:

$$P(firstcall)_1 = P(call)_1, \qquad\qquad (11.1)$$

$$P(call)_1 = P(\text{det})_1 P(occu)_1, \qquad\qquad (11.2)$$

where $P(call)_1$ = probability of calling the fire department at fire state 1,
$P(\text{det})_1$ = probability of fire detection at fire state 1,

$P(\text{occu})_1$ = probability of occupants calling the fire department at fire state 1, based on the consideration of the types of warning signals received by the occupants at various locations and the subsequent RIA process to make the call.

At state 2 of fire development, the probability of notifying the fire department for the very first time, $P(\textit{firstcall})_2$, is given by the following equations:

$$P(\textit{firstcall})_2 = [1 - P(\textit{firstcall})_1]P(\textit{call})_2, \tag{11.3}$$

$$P(\textit{call})_2 = P(A) + P(B) - P(A)P(B), \tag{11.4}$$

with

$$P(A) = P(\text{det})_2 P(\text{occu})_2$$

$$P(B) = P(\text{det})_2 P(\textit{staff})_2$$

where $P(\text{call})_2$ = probability of calling the fire department at fire state 2 by either occupants or security staff,

$P(\text{det})_2$ = probability of fire detection at fire state 2,

$P(\text{occu})_2$ = probability of occupants calling the fire department at fire state 2, based on the consideration of the types of warning signals received by the occupants at various locations and the subsequent RIA process to make the call,

$P(\text{staff})_2$ = probability of security staff calling the fire department at fire state 2, based on the consideration of the types of warning signals received by the staff and the procedure to make the call.

Note that in Equation 11.4, $P(\text{call})_2$ is based on the calling of the fire department by either the occupants or the security staff. Because these two calling events are not mutually exclusive, $P(\text{call})_2$ is equal to the union ($\cup$) of these two calling events.

At state 3 of fire development, the probability of notifying the fire department for the very first time, $P(\textit{firstcall})_3$, is given by the following equations;

$$P(\textit{firstcall})_3 = [1 - P(\textit{firstcall})_1 - P(\textit{firstcall})_2]P(\textit{call})_3, \tag{11.5}$$

$$P(\textit{call})_3 = P(C) + P(D) - P(C)P(D), \tag{11.6}$$

with

$$P(C) = P(\det)_3 P(occu)_3,$$

$$P(D) = P(\det)_3 P(direct)_3,$$

where $P(\text{call})_3$ = probability of calling the fire department at fire state 3,

$\quad P(\det)_3$ = probability of fire detection at fire state 3,

$\quad P(occu)_3$ = probability of occupants calling the fire department at fire state 3, based on the consideration of the types of warning signals received by the occupants at various locations and the subsequent RIA process to make the call,

$\quad P(direct)_3$ = probability of a direct link of fire detector activation to the fire department at fire state 3, if such an automatic notification is in place.

At states 4 and 5 of fire development, similar equations can be established to calculate the probabilities of notifying the fire department for the very first time at these later states of fire development.

11.2.1.2 Dispatch Time

Dispatch time is the time from the receipt of a fire notification in the fire department call centre (for example, 911 calls in Canada and the United States) to the dispatch of emergency responders (firefighters and firefighting apparatus) from appropriate emergency service providers (fire stations). The dispatch time is affected by the ability of the dispatchers to be able to quickly ascertain the location and severity of the fire incident. Their ability depends on their experience and the level of training they have received. The dispatch time is also affected by the volume of concurrent emergency calls. Too many concurrent calls can cause delays because not all of them can be attended to simultaneously by the dispatchers. Furthermore, the dispatch time is also affected by the availability of appropriate emergency service providers to respond. If they are not available because they have already gone to attend other emergencies, the dispatcher has to look for alternative emergency service providers to respond. Dispatch time, therefore, can vary from jurisdiction to jurisdiction. Table 11.2 shows, as an example, that the dispatch time in Canada and the United States is in the range of 0.5–2.0 minutes.

Table 11.2 Some examples of fire department response time based on the arrival of first responders.

Event time (min)	Canada (Gaskin and Yung, 1993)	United States U.S. Fire Administration National Fire Data Center, 2006)	United Kingdom (BSI, 2002, Part 5)
Dispatch time	1.0–1.5	0.5–2.0	–
Preparation time	0.5–1.0	–	–
Travel time	2.0–5.0	–	–
Response time	3.5–7.5[a]	2.0–8.0[b]	5.0–10.0[c]

[a]Represents fire departments with full-time firefighters only.
[b]Includes volunteer firefighters who may have to report to the fire station first; may or may not include dispatch time; covers the high-frequency response times and 77 % of all response times.
[c]Recommended response time limits for buildings located in areas with three different risk categories: 5 minutes for categories A (high risk) and B (medium risk) and 8–10 minutes for category C (low risk).

11.2.1.3 Preparation Time

Preparation time is the time from the receipt of a fire alert in a fire station to the time when the firefighters and their fire apparatus leave the fire station. Preparation time depends on how quickly the firefighters can assemble and get ready for an emergency response. Factors that can affect the preparation time include experience, training and whether the fire department is staffed with full-time or volunteer firefighters. Volunteer firefighters can take a longer time to prepare because some of them may have to go to the fire station first. Preparation time, therefore, varies from jurisdiction to jurisdiction. Table 11.2 shows, as an example, the preparation time in Canada for full-time firefighters is in the range of 0.5–1.0 minutes.

11.2.1.4 Travel Time

Travel time is the time from when the firefighters leave the fire station to the time when they arrive at the fire scene. Travel time depends primarily on the travel distance from the fire station to the fire scene. The more fire stations a jurisdiction has, the shorter is the travel distance and the travel time. Travel time also depends on the traffic conditions they encounter even though the fire trucks have emergency priority on the roads. If the primary route is blocked by heavy traffic, fire trucks have to use longer alternative routes. Travel time, therefore, varies from jurisdiction to jurisdiction. Table 11.2 shows, as an example, the travel time in Canada is in the range of 2.0–5.0 minutes.

11.2.1.5 Response Time

Response time, as mentioned earlier, is the sum of the *dispatch*, *preparation* and *travel* times. It is an important measure of how quickly a fire department can respond to a fire call. Response time is dependent on many factors and therefore can vary from jurisdiction to jurisdiction. Table 11.2 shows, as examples, the response times in Canada, United States and the United Kingdom, based on the arrival time of the first responders. The *response time* in Canada, representing fire departments with full-time firefighters, is in the range of 3.5–7.5 minutes. The *response time* in the United States, representing fire departments with both full-time and volunteer firefighters, is in the range of 2.0–8.0 minutes. The United States data, however, may or may not include the dispatch time which may explain why its response time can be as short as 2.0 minutes. The *response times* in the United Kingdom are not from actual data; they are instead the recommended upper limits for fire service intervention by the British Standards (BSI, 2002, Part 5). The recommended upper limit is 5 minutes for buildings in 'Category A' area (high risk); 5 minutes for 'Category B' area (medium risk) and 8–10 minutes for 'Category C' area (low risk). The recommended upper limit of 5 minutes in the United Kingdom for high and medium risk areas is similar to the average response time, about 5 minutes, in Canada and the United States.

11.2.1.6 Setup Time

Setup time is the time from when the firefighters arrive at the fire scene to the time when they commence occupant rescue and fire extinguishment efforts. Setup time depends on many factors. These include the height of the building, the locations of the hydrants and the hook-up time of the standpipes. Setup time, therefore, can vary from jurisdiction to jurisdiction. A study in Canada suggests that the *setup time* is in the range of 3–7 minutes (Gaskin and Yung, 1993).

11.2.1.7 Intervention Time

Intervention time, as mentioned earlier, is the sum of the notification, response and setup times. It is the time from ignition to the time when firefighters commence occupant rescue and fire extinguishment efforts. As such, it is an important measure of how quickly firefighters can intervene at a fire development. A quick intervention time allows the firefighters to control a fire before it can develop into a severe fire; whereas a long intervention time allows the fire to develop and cause harm to the occupants and property before the firefighters can control it.

The *intervention time* can be estimated based on the various event times discussed above. For example, the quickest notification time, based on the activation times of smoke detectors, is in the range of 0.5–1.8 minutes. It can be much longer if there are no smoke or heat detectors and the notification is made by people who happen to notice the fire and take the trouble to call the fire department. The *response time*, based on the limited data shown in Table 11.2, is in the range of 3.5–8.0 minutes. The *setup time*, based on the Canadian study, is in the range of 3.0–7.0 minutes. The *intervention time*, based on all these limited data, is in the range of 7.0–16.8 minutes.

The above discussion shows that the *intervention time* can be easily over 10 minutes. The characteristic time of fire development in a compartment to a flashover fire, on the other hand, is usually much less than 10 minutes (see Chapter 7). Flashover fires are severe fires that can generate significant amounts of heat and smoke in a short time. The resultant smoke spread can create untenable conditions in the building quickly, blocking evacuation routes and posing hazards to the occupants. The high heats that are generated in flashover fires can also breach boundary elements. The resultant fire spread can cause significant damages to the building and property.

The risk of life-loss to the occupants depends on whether they can evacuate before the arrival of the critical smoke conditions in the evacuation routes that prevent evacuation. For those who cannot evacuate in time and are trapped in the building, their lives depend on the length of their exposure to untenable conditions. The length of exposure depends on the characteristic time of fire development to a flashover fire and the intervention time when occupant rescue effort is commenced by the fire department. The longer is the intervention time from the characteristic time of fire development, the longer is the exposure and the larger is the risk of life-loss to the trapped occupants.

Similarly, the risk of property loss depends on the level of fire resistant construction. Proper fire resistant construction can help minimize the extent of fire spread. The extent of fire spread also depends on the characteristic time of fire development to a flashover fire and the intervention time when fire extinguishment effort is commenced by the firefighters. The longer the intervention time from the characteristic time of fire development, the larger is the extent of fire spread and the larger the risk of property loss.

The assessment of risks to occupants and property will be discussed in more detail later in this chapter.

11.2.2 Response Resources

The status of a fire development when the firefighters arrive at the fire scene depends on the response time, as was discussed in the last section. The quicker the response time, the less severe is the fire, and the smaller the effort that is required to fight the fire and to rescue any trapped occupants. The level of effort that the firefighters can provide, on the other hand, depends on the level of resources that is dispatched and available at the fire scene. The effectiveness of the occupant rescue and fire extinguishment efforts, therefore, depends on a quick response time and the availability of adequate resources.

A number of fire department computer models have been developed in recent years which can be used to simulate the effectiveness of occupant rescue and fire extinguishment efforts based on the response time and available resources. Some of these models are standalone models, such as the Australian Fire Brigade Intervention Model (*FBIM*) which guides a user to use a series of charts to assess the time lines and effectiveness of occupant rescue and fire extinguishment efforts (Merchant, Kurban and Wise, 2001). Others are part of larger fire risk assessment models, such as the Fire Department Response and Effectiveness sub-models that are part of the Canadian Fire Risk Evaluation and Cost Assessment Model (FiRECAM) (Yung, Hadjisophocleous and Proulx, 1997). Assuming that the firefighting equipment is adequate, the resources that have a direct impact on the effectiveness of the occupant rescue and fire extinguishment efforts are mainly the crew size and water resource.

11.2.2.1 Crew Size

The crew size affects both the occupant rescue and fire extinguishment efforts. The more firefighters are dispatched to the fire scene, relative to the number of trapped occupants, the more effective is the rescue effort. Also, the more firefighters are dispatched to the fire scene, relative to the size of the fire, the more effective is the fire extinguishment effort to contain and the fire.

11.2.2.2 Water Resource

The ability to extinguish a fire in a building depends on the availability of adequate water resource to combat the fire. The required water flow rate depends on the intensity of the fire at the time when the firefighters commence fire extinguishment effort. The required water flow rate can be assessed by the following heat balance equation (Torvi, Hadjisophocleous and Guenther, 2001).

$$RFL_w = \frac{\dot{Q}}{\eta_w L_w},\tag{11.7}$$

where RFL_w is the required water flow rate (litre per second) to absorb the heat release rate of the fire, $\dot{Q}$ is the heat release rate of the fire (MW), L_w is the latent heat of vapourization of water ($2.26\,\text{MW·L}^{-1}\text{·s}^{-1}$), and η_w is the efficiency of water application by the firefighters to absorb the heat from the fire. The water application efficiency η_w has a value in the range of 0.1–0.3 (Torvi, Hadjisophocleous and Guenther, 2001). The firefighting effectiveness drops if the water resource is less than the RFL_w. A 10 MW fire would require a water flow rate of 14.7–44.2 L s^{-1} (234–701 US gpm). Water resource is usually adequate in urban areas, but may be limited in rural areas.

11.2.3 Occupant Rescue and Fire Extinguishment Effectiveness

The firefighter's effectiveness to rescue trapped occupants and to extinguish the fire is affected by how quickly the *intervention time* is and how adequate is the available rescue and fire extinguishment resources. The quicker is the *intervention time*, the less severe is the smoke spread, the lesser is the number of trapped occupants, and, with adequate resources, the greater is the *rescue effectiveness*. Similarly, the quicker is the *intervention time*, the less severe is the fire, and, with adequate resources, the greater is the *fire extinguishment effectiveness*.

The effectiveness of firefighter's occupant rescue effort can be modelled as a function of the ratio of the number of trapped occupants to the firefighter's crew size (Benichou, Yung and Hadjisophocleous, 1999). The larger is the ratio, the larger is the number of trapped occupants, and the less effective is the rescue effort. Figure 11.2 shows their modelling of the exponential drop in rescue effectiveness P_{RES} with the increase in the ratio of the trapped occupants to the fire-fighter's crew size R_{TC}. This figure illustrates the importance of early occupant evacuation so that the number of trapped occupants can be minimized.

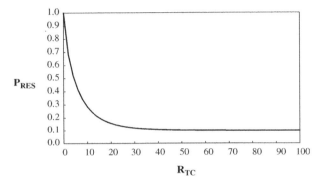

Figure 11.2 Occupant rescue effectiveness P_{RES} decreases with the increase in the ratio of the trapped occupants to the firefighter's crew size R_{TC} (from Benichou, Yung and Hadjisophocleous, 1999, reproduced by permission of Interscience Communications Ltd).

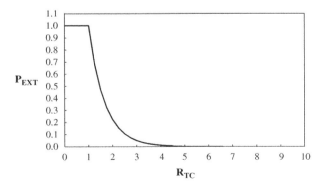

Figure 11.3 Fire extinguishment effectiveness P_{EXT} begins to decrease when the firefighter's intervention time exceeds the flashover time, and continues to drop with the increase in the ratio of the intervention time to the flashover time R_{FO} (from Benichou, Yung and Hadjisophocleous, 1999, reproduced by permission of Interscience Communications Ltd).

The effectiveness of fire extinguishment can be modelled as a function of the ratio of the firefighter's intervention time to the flashover time (Benichou, Yung and Hadjisophocleous, 1999). The larger is the ratio, the more intense is the fire, and the less effective is the fire extinguishment effort. Figure 11.3 shows their modelling of the exponential drop in the fire extinguishment effectiveness P_{EXT} with the increase in the ratio of the firefighter's intervention time to the flashover time R_{FO}. This figure illustrates the importance of quick intervention time so that the intensity of the fire to be extinguished is not as great.

11.3 Occupant Fatality and Property Loss Modelling

In the previous section, the fire department *intervention time* and *response resources* were discussed. The intervention time relates to the status of the fire development at the time when the firefighters commence their occupant rescue and fire extinguishment efforts; whereas the response resources relate to the effectiveness of their occupant rescue and fire extinguishment efforts. In this section, we will discuss how the occupant fatalities and property loss can be assessed. They are assessed based on the length of exposure to untenable conditions from fire ignition to the time of fire department intervention and the effectiveness of their occupant rescue and fire extinguishment efforts.

11.3.1 Occupant Fatality

Occupant fatalities can be assessed based on the exposure to toxic gases and high temperature from the smoke and fire spread in the building. The occupants who are at risk are those who can not evacuate in time and are trapped in the building (see Chapter 10).

The harmful effect of toxic gases to people is determined based on research in toxicology. The science is complex which involves the use of animals and the extrapolation of the animal data to humans. In addition, every human body is different and therefore the effect on each person is different. Furthermore, various toxic gases are generated from fires, depending on the materials that are burned. The combined effect of multiple toxic gases is even more complex. Nevertheless, there are simple models that can be used to provide some assessments of the harmful effect of toxic gases (Hadjisophocleous and Yung, 1992; Yung and Benichou, 2002). These models assess occupant incapacitation and deaths based on the asphyxiant effect of carbon monoxide (CO) and the hyperventilating effect of carbon dioxide (CO_2). Both of these gases are always present in fires. CO causes asphyxiation by combining with blood's hemoglobin to form carboxyhemoglobin (COHb) which reduces blood's ability to carry oxygen; whereas CO_2 stimulates hyperventilation which causes higher intake of CO. Incapacitation occurs when COHb reaches 30 % (Purser, 2002). Death soon follows with further intake of CO and further increase in COHb.

The asphyxiant effect of CO can be calculated using the following equation which compares the intake of CO over time to the critical 30 % COHb (Purser, 2002):

$$FID_{CO} = \frac{K}{D} \int_0^t CO^{1.036} \, dt, \tag{11.8}$$

where FID_{CO} is the fraction of incapacitating dose from CO, K is a constant which is 8.2925×10^{-4} for light activity (breathing rate at 25 L min^{-1}), CO is the CO concentration (ppm), D is the carboxyhemoglobin concentration at incapacitation which is 30 for 30 %, and t is the time (minutes) from ignition to the time of fire department intervention. The CO concentration at any time and at any location in a building is determined from smoke spread calculations (see Chapter 9).

The hyperventilating effect of CO_2 is to increase FID_{CO} as a result of higher breathing rate. The multiplication factor can be calculated using the following equation (Purser, 2002):

$$VCO_2 = \frac{\exp(0.2496 \, CO_2 + 1.9086)}{6.8}, \tag{11.9}$$

where VCO_2 is a multiplication factor for CO_2 induced hyperventilation and CO_2 is the CO_2 concentration (%). The CO_2 concentration at any time and at any location in a building is determined from smoke spread calculations (see Chapter 9).

By combining Equations 11.2 and 11.3, the accumulated *fractional incapacitating dose FID* can be obtained by integration with time, as:

$$FID = \frac{K}{D} \int_0^t CO^{1.036} \, VCO_2 \, dt \tag{11.10}$$

In the above equation, FID is limited to a maximum value of 1 (when incapacitation occurs) even though the equation may produce a value higher than 1. If the gas concentrations are constant, Equation 11.10 can be easily plotted to show what the incapacitation times are for various

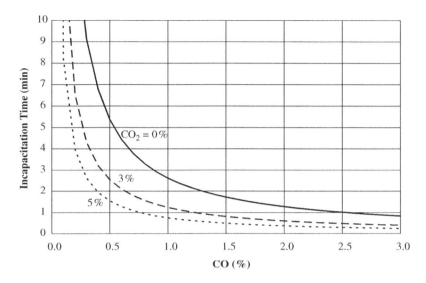

Figure 11.4 Incapacitation time decreases with the increase in CO and CO_2 concentrations.

CO and CO_2 concentrations. Figure 11.4 shows, as expected, that the incapacitation time decreases with the increase in CO concentration and also with the increase in CO_2 concentration. Note that the concentration values of CO and CO_2 that are plotted in Figure 11.4 are typical compartment fire values (see Chapter 7). When the CO concentration reaches 0.5 %, the incapacitation time is only 5 minutes even with a 0 % CO_2 concentration. With a higher CO_2 concentration, the incapacitation time is much less. This shows why it is important to have a quick evacuation time so that the occupants can get to a safe place before the spread of the toxic smoke in the building. This also shows why it is important to have a quick fire department intervention time so that those occupants who can not evacuate in time and are trapped in the building can be rescued before they are harmed by the toxic smoke.

Incapacitation can also occur as a result of high temperature; the tolerance time of naked skin drops quickly when the temperature is above 100 °C (Purser, 2002). The modelling of the probability of incapacitation from high temperatures is very complex. Nevertheless, there are simple models that can be used to provide some assessments of the harmful effect of high temperature. The following is one example which is based on the assumption that the incapacitation has a value of 1 when the hot gas temperature reaches 100 °C (Hadjisophocleous and

Yung, 1992; Yung and Benichou, 2002):

$$PIT = \frac{T_t - T_a}{100 - T_a}, \tag{11.11}$$

where PIT is the probability of incapacitation from temperature, $T_t - T_a$ is the temperature rise (°C), and T_a is the initial building temperature (°C). Similar to FID, PIT is also limited to a maximum value of 1 (when incapacitation occurs) even though the equation may produce a value higher than 1.

The risks to the occupants from toxic gases, FID, and from temperature, PIT, are two independent and not mutually exclusive events. Therefore, the probability of smoke hazard P_{SS} can be calculated as the union (∪) of the two events:

$$P_{SS} = FID + PIT - (FID \cdot PIT). \tag{11.12}$$

The number of occupants killed by the smoke hazard in each location and at the fire department's intervention time can be calculated based on the number of trapped occupants and the smoke hazard at that location and at the intervention time:

$$OF_{SS} = OCC_{TR}P_{SS}, \tag{11.13}$$

where OF_{SS} is the number of *occupant fatalities* as a result of smoke hazard in a particular location and at the intervention time, OCC_{TR} is the number of trapped occupants in that location at the intervention time, P_{SS} is the *smoke hazard* in that location at the intervention time. (See Chapter 9 and Chapter 10, for the discussions of smoke spread and the trapped occupants.) Those who are not killed by the smoke hazard depend on the firefighter's occupant rescue effectiveness. The number of occupants who are not killed by the smoke hazard and can not be rescued by the firefighters in that location and at the intervention time is:

$$OF_{NR} = OCC_{TR}(1 - P_{SS})(1 - P_{RES}), \tag{11.14}$$

where OF_{NR} is the additional number of occupant fatalities as a result of no rescue by the firefighters in a particular location and at the intervention time, P_{RES} is the firefighter's occupant rescue effectiveness at the intervention time. The trapped occupants who are not rescued are presumed to be killed by further smoke spread and also by the eventual fire spread (see Chapter 8). The total number of occupant fatalities in a particular location and at the intervention time as a result of both smoke hazard and ineffective occupant rescue effort is therefore:

$$OF = OCC_{TR}[P_{SS} + (1 - P_{SS})(1 - P_{RES})]. \tag{11.15}$$

The assessment of the total number of deaths for a particular fire requires the repeat use of Equation 11.15 to calculate the number of fatalities in all locations and for all fire scenarios. The use of computer models can help with this effort.

Note that Equation 11.15 shows the importance of early evacuation to minimize the number of trapped occupants OCC_{TR}, the use of smoke control system to minimize the smoke hazard P_{SS}, and the importance of early notification to the fire department to increase the firefighter's occupant rescue effectiveness P_{RES}.

11.3.2 Property Loss

Property loss can be assessed, similar to the assessment of occupant fatalities, based on smoke spread and firefighter's fire extinguishment effectiveness. Equation 11.15 can be re-written for property values as:

$$PL = PRO[P_{SS} + (1 - P_{SS})(1 - P_{EXT})], \tag{11.16}$$

where PL is the *property loss* due to both smoke hazard and ineffective fire extinguishment effort in a particular location and at the intervention time, PRO is the property value in that location and at the intervention time, P_{SS} is the smoke hazard in that location at the intervention time, P_{EXT} is the firefighter's fire extinguishment effectiveness at the intervention time. The property value PRO is the property value that is at risk, which include both structural and content. The property value that is not saved by fire extinguishment is presumed to be lost by further smoke spread and also by the eventual fire spread (see Chapter 8).

Similar to the assessment of total occupant fatalities for a particular fire, the assessment of the total property loss for a particular fire requires the repeat use of Equation 11.16 to calculate the property loss in all locations and for all fire scenarios. The use of computer model can help with this effort.

Again, Equation 11.16 shows the importance of the use of smoke control system to minimize the smoke hazard P_{SS}, and the importance of early notification to the fire department to increase the firefighter's fire extinguishment effectiveness P_{EXT}.

11.4 Fire Protection Measures to Provide Effective Occupant Rescue and Fire Extinguishment Efforts

Quick response time and effective occupant rescue and fire extinguishment efforts by the fire department, as was discussed in previous sections, can help reduce occupant fatalities and property loss. There are many ways to help facilitate quick response time and effective occupant rescue and fire extinguishment efforts. The following are three examples of such fire protection measures.

11.4.1 Automatic Notification System to Provide Early Notification to Fire Department

Fire department's *intervention time*, as was discussed in detail in the previous sections, is the cumulated sum of many event times. One event time that is most unpredictable is the *notification time*. The notification time to the fire department can be quite short or quite long, depending on whether or not there are security staffs present or automatic notification systems in place. If there is an on-site security staff who would notify the fire department immediately upon fire detector activation, or there is an automatic notification system in place such as a direct link of fire detector activation to the fire department, the notification time can be quite close to the fire detection time. If there are no such protection systems in place and the notification is dependent entirely on occupants who have noticed the fire and called the *fire department*, the notification time can be quite long. Therefore, one way to provide early notification to the fire department is to provide the building with a security staff or an automatic notification system. If such protection systems can not be provided, the occupants should be trained to learn to call the fire department immediately upon receiving fire warnings.

11.4.2 Adequate Number of Fire Stations to Provide Quick Response Time

An event time that can affect fire department's *response time* is the *travel time*. Travel time depends on the distribution and the number of fire stations in a certain area. To provide a certain minimum travel time, a proper distribution and an adequate number of fire stations in a certain area is needed. In urban areas, fire departments usually have adequate number of fire stations to provide a certain minimum travel time. In rural areas, however, they may not be able to provide the same number of fire stations. Buildings in rural areas, therefore, may require different fire protection strategies.

11.4.3 Adequate Resources to Provide Effective Occupant Rescue and Fire Extinguishment Efforts

The effectiveness of occupant rescue and fire extinguishment efforts depends on the availability of adequate resources, such as crew size, proper rescue and firefighting equipment and water resources. As was discussed in previous sections, the fire department's effectiveness would drop if there are insufficient resources. Fire departments usually have adequate resources for urban areas, but not necessarily in rural areas. Buildings in rural areas, therefore, may require different fire protection strategies.

Figure 11.5 is a photograph of the University of Berkeley's helmet-mounted 'FireEye' device that can help firefighters track their current position superimposed on a floor map of the building. It is shown here just as an example of research and development efforts worldwide to come up with better fire rescue and fighting equipment.

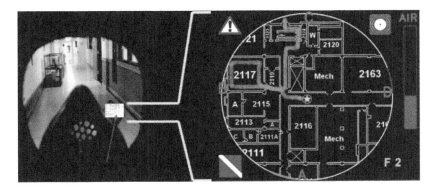

Figure 11.5 The University of Berkeley's helmet-mounted 'FireEye' device can help firefighters track their current position superimposed on a floor map of the building (from Dill, 2007, reproduced by permission of Joel Wilson and the Berkeley Science Review).

11.4.4 Fire Department Response Scenarios

The probability of success or failure of fire protection measures can be used to create a set of fire department response scenarios for fire risk assessments. We will consider here only the three aforementioned fire protection measures that can be implemented: automatic notification system, adequate fire stations and adequate crew and firefighting resources.

Each of the first two chosen fire protection measures can help shorten the *intervention time*. Automatic notification system can help provide early *notification time*; whereas adequate number of fire stations can help provide quick *travel time*. Adequate crew and firefighting resources would provide effective occupant rescue and fire extinguishment efforts. Quick response time and adequate resources would help to minimize the length of exposure of any trapped occupants to untenable fire and smoke conditions before they are rescued, and would also help minimize the extent of fire and smoke spread in the building before they are extinguished.

Figure 11.6 shows, based on the success and failure of the above three selected fire protection measures, a total of eight possible fire department response scenarios. Each of these scenarios has an implied intervention time and available resources. Scenario A, with all the fire protection measures successfully operating, has the shortest intervention time and adequate resources. Scenario H, at the other extreme with all the fire protection measures fail, has the longest intervention time and inadequate resources.

In Figure 11.6, the probabilities of success and failure of automatic notification systems by direct link of fire detector activation to fire department depend on the design and maintenance of these systems. The probabilities of success and failure of immediate notification by security staff depend on the training of the security staff. The probabilities of success and failure of adequate number of fire stations and adequate rescue and firefighting resources depend on the availability of these resources and the level of demand for their service. These probability values should be agreed upon by fire safety engineers and authorities having jurisdiction. Reliability and effectiveness of fire protection systems will be discussed in more detail in Chapter 13.

11.5 Summary

Occupants who are trapped in a building because they can not evacuate in time before the arrival of the untenable fire and smoke conditions

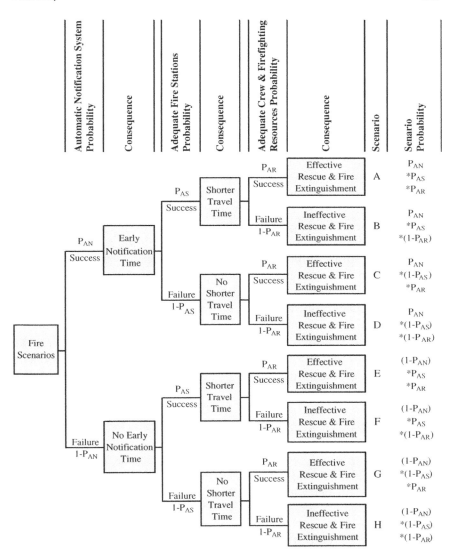

Figure 11.6 Fire department response scenarios based on probabilities of success or failure of fire protection measures. *Note:* P_{AN} = probability of success of automatic notification system, P_{AS} = probability of success of adequate fire stations, P_{AR} = probability of success of adequate crew and firefighting resources.

face the risk of losing their lives unless the fire department can respond and rescue them in time. The risk of life-loss to the trapped occupants depends on the length of their exposure to the untenable fire and smoke conditions before they are rescued by the fire department. Similarly, the risk of property loss to the building depends on the extent of fire and smoke spread in the building before they are extinguished by the fire department.

The time when the fire department can commence the occupant rescue and fire extinguishment efforts is called the *intervention time*. Occupant fatalities and property loss are assessed based on the length of exposure to untenable fire and smoke conditions up to the intervention time when firefighters arrive and commence rescue and firefighting efforts. The effectiveness of firefighter's occupant rescue and fire extinguishment efforts depend on dispatched crew size and firefighting resources, which include firefighting equipment and water resources.

Fire protection measures that can help provide early intervention time and effective occupant rescue and fire extinguishment efforts include the use of on-site security staffs and automatic notification systems to provide early notification; adequate distribution of fire stations to provide quick travel time; and adequate crew and firefighting equipment resources to provide effective occupant rescue and fire extinguishment efforts. An automatic notification system can be a direct link of fire detector activation to the fire department.

Based on the success or failure of these fire protection measures, fire department response scenarios can be constructed. The probabilities of the fire department response scenarios, together with the calculation of the firefighter's intervention time and their rescue and firefighting effectiveness, can be used to assess the risk of life-loss to the occupants and the risk of property loss to the building.

11.6 Review Questions

11.6.1 Calculate the probabilities of notification at different states of fire development, $P(\text{firstcall})_i$. Assume the probability of detection at each state, $P(\text{det})_i$, is 0.9; the probability of occupants calling the fire department at each state, $P(\text{occu})_i$, is 0.2; the probability of notification by security staff at state 2, $P(\text{staff})_2$, is 0.9; and the probability by direct link to the fire department at state 3, $P(\text{direct})_2$ is 0.9.

11.6.2 Repeat the same calculations in the above question by assuming no security staff and no automatic notification system and compare the two.

References

Benichou, N., Yung, D. and Hadjisophocleous, G.V. (1999) Impact of Fire Department Response and Mandatory Sprinkler Protection on Life Risks in Residential Communities, Proceedings of Interflam '99, July 1999, Edinburgh, Scotland, pp. 521–32.

BSI (2002), Fire service intervention, Part 5, *Application of Fire Safety Engineering Principles to the Design of Buildings*, PD7974-5, British Standards Institution, London, UK, p. 11.

Dill, J. (2007) Fighting Fire with F.I.R.E. – New technology Improves First Responder Communication, *Berkeley Science Review*, (12), Spring, 18–19.

Gaskin, J. and Yung, D. (1993) *Canadian and U.S.A. Fire Statistics for Use in the Risk-Cost Assessment Model*, IRC Internal Report No. 637, National Research Council Canada, January 1993, Ottawa, Canada, 18 pages.

Hadjisophocleous, G.V. and Yung, D. (1992) A model for calculating the probabilities of smoke hazard from fires in multi-storey buildings. *Journal of Fire Protection Engineering*, 4(2), 67–80.

Hadjisophocleous, G.V. and Yung, D. (1994) Parametric Study of the NRCC Fire Risk-Cost Assessment Model for Apartment and Office Buildings, Proceedings of the Fourth International Symposium on Fire Safety Science, June 1994, Ottawa, Canada, 829–40.

Merchant, R., Kurban, N. and Wise, S. (2001) Development and application of the fire brigade intervention model. *Fire Technology*, 37(3), 263–78.

Proulx, G. and Hadjisophocleous, G.V. (1994) Occupant Response Model: A Sub-Model for the NRCC Risk-Cost Assessment Model, Proceedings of the Fourth International Symposium on Fire Safety Science, June 1994, Ottawa, Canada, pp. 841–52.

Purser, D.A. (2002) Toxicity assessment of combustion products, *SFPE Handbook of Fire Protection Engineering 2002*, 3rd edn, Section 2: Chapter 6, National Fire Protection Association, Quincy, MA.

Torvi, D.A., Hadjisophocleous, G.V. and Guenther, M.B. (2001) Estimating water requirements for firefighting operations using FIERAsystem. *Fire Technology*, 37(3), 235–62.

U.S. Fire Administration/National Fire Data Center (2006) Structure fire response times. *Topical Fire Research Series*, 5(7), August 2006, 5 pages.

Yung, D. and Benichou, N. (2002) How design fires can be used in fire hazard analysis. *Fire Technology*, 38(3), 231–42.

Yung, D., Hadjisophocleous, G.V. and Proulx, G. (1997) Modelling Concepts for the Risk-Cost Assessment Model FiRECAM and Its Application to a Canadian Government Office Building, Proceedings of the Fifth International Symposium on Fire Safety Science, March 1997, Melbourne, Australia, pp. 619–30.

12

Uncertainty Considerations

12.1 Overview

How to consider uncertainties, especially those related to fire safety designs, is discussed in this chapter. There is uncertainty in all disciplines and in the case of fire safety, uncertainties in design parameters may produce uncertainties in fire safety designs and in fire risk assessment. Many factors can lead to fire safety parameters with uncertain values, but these parameters can be generally grouped as follows:

1. parameters with a probability distribution of different values,
2. parameters with random values,
3. parameters with unknown values and
4. parameters with future unknown values.

The probability of success or failure in fire safety designs for parameters with a probability distribution, can be analysed in a multi-dimensional space with the space divided by the fire safety design into safe and unsafe regions. The following two techniques are discussed that can be used to find the cumulative probability in the safe region:

1. The Monte Carlo method which employs repeated random sampling to obtain the cumulative probability in the safe region.
2. The β reliability index method which uses the largest hyper-sphere that can fit into the safe region to determine the cumulative probability in the safe region.

There are no standard treatment methods for parameters with uncertain values but without a known probability distribution. The uncertain values may be determined through research or survey. Otherwise the uncertain values need to be assumed and agreed upon between fire safety engineers and regulators. Conducting a parametric study is one method to check what value can be assumed.

12.2 What Are the Uncertainties?

Uncertainty exists in fire risk assessment because we do not know for certain the values of many fire safety parameters. For example, we may not know for certain the values of some parameters that we input into fire or human behaviour models. We may also not know for certain that the modelling equations that we employ in fire or human behaviour models are accurate. Uncertainty in either input values or modelling equations can have an impact on the accuracy of our predictions of fire growth and human behaviour and eventually our assessment of fire risks. In this chapter, we will discuss mainly the uncertainty in parameter values that we employ in fire risk assessment. The uncertainty in modelling equations is a more fundamental issue that is being addressed continuously by the fire research community to make these equations more accurate.

The issue of uncertainty is not unique in fire risk assessment. It is an issue that exists in many disciplines, from quantum mechanics to financial markets. A good reference from the perspective of fire engineering is the chapter on uncertainty in the SFPE Handbook (Notarianni, 2002). Uncertainty in parameter values can be a result of many factors, but can be generally grouped into the following four major types.

12.2.1 Parameters with a Probability Distribution of Different Values

These are parameters that do not have a single value for a particular condition. Instead, they can take on many values for a particular condition and hence there is an uncertainty about their values. However, each of their values can usually be identified with a certain probability of occurrence. The uncertainty of these values, therefore, can be addressed through the *probability distribution* of their values. An example of this type of uncertainty is the fuel load density which was discussed previously in Chapter 8. The probability distribution of the fuel load density was used to assess the probability of failure of the boundary element with a certain design fire resistance rating (FRR).

12.2.2 Parameters with Random Values

These are also parameters that do not have a single value for a particular condition. Instead, they can take on many random values for a particular condition and hence there is an uncertainty about their values. Random values are values that can be any value and each value can not be easily identified with a certain probability of occurrence. As a result of the large number of possible values and the lack of association of probability of occurrence to each value, the uncertainty of *random parameters* can not be easily addressed. Only in some special cases when the random parameters can be found to give rise to certain statistical outcomes can the uncertainty of these parameters be addressed. An example of this is the random parameters of the ignition point and the fuel arrangement in the compartment of fire origin which were discussed previously in Chapter 7. The ignition point can be any point in the compartment of fire origin and the fuel arrangement can have infinite variations. The uncertainty of these two parameters can not be easily addressed because of the large number of possible variations and the lack of probability association. However, their random nature was argued to be the cause for giving rise to three different fire growth scenarios: smouldering, non-flashover and flashover fires. Each of these fire growth scenarios shows up in fire statistics with a certain probability of occurrence. The uncertainty of the ignition point and the fuel arrangement, therefore, was addressed through the probabilities of occurrence of the three resultant fire scenarios, rather than the random values of these two parameters.

12.2.3 Parameters with Unknown Values

These are parameters with values that are unknown or yet to be determined. Usually, their values have to be assumed and hence the uncertainty. One example is the value of the parameter of 'no smoking material' on the probability of fire occurrence in an apartment unit, which was discussed previously in Chapter 5. There is no statistical information that can be easily found on the probability of success or failure of implementing a 'no smoking material' plan so that there will be a lower rate of fire occurrence. Until such value can be determined, the value has to be assumed. In fire risk assessment, any assumed value needs to be agreed upon between the fire safety engineer and the regulators.

12.2.4 Parameters with Future Unknown Values

These are parameters with values that may be known at the present time but may change in the future. The potential change in value in the future will have an impact on the level of fire risk in the future and hence the uncertainty. One example of such parameters with future unknown values is the flammability properties of furniture materials, which were discussed previously in Chapter 7. The furniture materials can change in the future and hence their flammability properties. The level of fire risk in the future will be different as a result of these changed values. This is one reason why fire risk assessment needs to be reviewed from time to time in order to take into account possible changes that may affect the level of fire risk in the future.

12.3 Treatment of Uncertainty

In this section, we will discuss the treatment of uncertainty in fire risk assessment. We will discuss mainly the treatment of uncertainty in parameter values that affects the probability of success or failure of fire safety designs which in turn affects the level of fire risk. First, we will discuss the treatment of parameters with a probability distribution of different values. Then, we will discuss the general treatment of parameters with uncertain values.

12.3.1 Treatment of Parameters with a Probability Distribution of Different Values

The treatment of parameters with a probability distribution of different values was discussed previously in Chapter 8. In that earlier discussion, only two design parameters were considered. The first parameter, the equivalent standard fire time, was considered to have a probability distribution (see Figure 8.3 in Chapter 8). The second parameter, the design FRR, was assumed to have a fixed value with no probability distribution of different values. In that simple case, the probability of success in resisting the equivalent standard fire was represented by the area under the probability distribution curve below the design FRR (the safe region, see Figure 8.3 in Chapter 8). Similarly, the probability of failure in resisting the equivalent standard fire was represented by the area under the probability distribution above the design FRR (the unsafe region).

If the fire safety design involves more than one parameter with a probability distribution, the determination of the probability of design success (the *safe region*) and design failure (the *unsafe region*) can be assessed by following the same methodology that was employed in the previous simple case. Instead of finding the areas under a curve that represent the safe and unsafe regions, the methodology is to find the spaces in a multi-dimensional space that represent the safe and unsafe regions. For example, a design problem can be written as a function of n controlling parameters $(x_1$ to $x_n)$:

$$G(x_1, x_2, \ldots, x_n) = 0. \tag{12.1}$$

The function can be constructed such that when G has values that are smaller or equal to 0, the design is safe; and when G has values that are greater than 0, the design is unsafe. If this function is plotted in a *multi-dimensional space*, as depicted in Figure 12.1, the function becomes a *hypersurface*. The design hypersurface divides the multi-dimensional space into two regions: the safe region, represented by values of G that are either smaller or equal to 0; and the unsafe region, represented by values of G that are greater than 0.

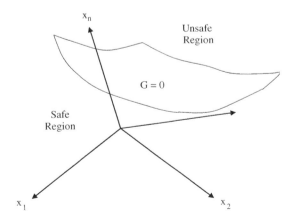

Figure 12.1 Design hypersurface divides a multi-dimensional space into a safe and an unsafe region, with the values of G in the safe region equal to or smaller than 0, and those in the unsafe region greater than 0.

In the multi-dimensional space, each point has a G value and an associated probability value. The probability value is the combined value of all the controlling parameter probability values at that point. Since these controlling parameters are not mutually exclusive, the combined probability value at each point is the product of the probability values of all the controlling parameters, as described by the following equation:

$$p(G) = p(x_1)\,p(x_2)\ldots p(x_n).\qquad (12.2)$$

The probability that the design is safe is represented by the cumulative probability in the safe region. Conversely, the probability that the design is not safe is represented by the cumulative probability in the unsafe region. We will discuss in the following some of the methods that can be used to determine this cumulative probability.

12.3.1.1 Monte Carlo Method

The *Monte Carlo method* is a computational method that employs repeated random sampling to obtain approximate solutions to multi-dimensional problems that has no exact deterministic solutions. It was developed by scientists at Los Alamos in the 1940s for application to atomic physics problems. The name 'Monte Carlo' was coined by the scientists because the method resembled playing Russian roulette in the casinos in Monte Carlo. With this method, the larger the sampling, the more accurate is the approximation, regardless of the number of dimensions. The error is scaled by $1/\sqrt{N}$, where N is the number of sampling. However, the larger the sampling, the larger is the computational time.

To apply the method to the present problem, random points in the multi-dimensional space are generated and those that fall within the safe region are counted. For each random point, the combined probability based on Equation 12.2 is computed. The cumulative probability in the safe region is obtained by repeated sampling. The cumulative probability in the unsafe region is the complementary value which is equal to 1 minus the cumulative probability in the safe region. The size of the sampling that should be used depends on how small the error is desired. Since the error is scaled by the square root of the sampling size, each four-time increase in the sampling size cuts the error by half.

12.3.1.2 β Reliability Index Method

The *β reliability index* method is a method that has been used in structural engineering for decades. The β index gives an indication of the level of reliability, or certainty, that the structural design is safe. It acts as a measuring stick of the cumulative probability in the safe region in Figure 12.1. The higher the β index value, the higher is the probability that the structural design is safe. The β reliability index method was first applied to fire safety engineering by Frantzich *et al.* in 1997 (Frantzich *et al.*, 1997).

To derive the β reliability index, the first order second moment (*FOSM*) method is used (Hasofer and Lind, 1974). The FOSM method uses first order approximation and standard deviations (second moment) in its approach and hence the name. It assumes that the probability distributions of the parameters are well behaved, such as those that follow a normal distribution $N(\mu, \sigma)$ with a mean value, μ, and a standard deviation, σ. This allows the β reliability index to be determined based entirely on the mean values and standard deviations without the probability distributions themselves appearing in the solutions.

Figure 12.2 is a plot of a normal probability density distribution for a parameter, x_i, with a mean value, μ_i and a standard deviation, σ_i. The equation of the normal probability density distribution, $N(\mu_i, \sigma_i)$, for a parameter, x_i, is:

$$p(x_i) = \frac{1}{\sqrt{2\pi}\,\sigma_i}\,e^{-\frac{(x_i - \mu_i)^2}{2\sigma_i^2}}. \tag{12.3}$$

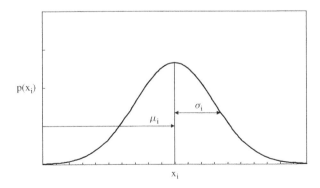

Figure 12.2 Normal probability density distribution for a parameter, x_i, with a mean value, μ_i and a standard deviation, σ_i.

The above normal probability density distribution yields a cumulative probability of 1 when integrated from $-\infty$ to $+\infty$. In Figure 12.1, the combined probability density at any point is given by Equation 12.2, with the individual probability density given by Equation 12.3. The integration of the combined probability density in the safe region gives the cumulative probability in the safe region, which represents the probability of success that the design is safe. The integration in the safe region is represented by the following equation:

$$p(safe\ region) = \frac{1}{(\sqrt{2\pi})^n} \iiint e^{-\frac{1}{2} \sum_{1}^{n} \frac{(x_i - \mu_i)^2}{\sigma_i^2}} d\frac{x_1}{\sigma_1} d\frac{x_2}{\sigma_2} \cdots d\frac{x_n}{\sigma_n}. \qquad (12.4)$$

The above integration can be simplified using the *Hasofer–Lind trans-formation* (Hasofer and Lind, 1974):

$$u_i = \frac{x_i - \mu_i}{\sigma_i}; \quad \text{with } i = 1, 2, \ldots, n. \qquad (12.5)$$

The transformation moves the integration from a physical space to a standard space where all transformed parameters have a zero mean and a unit standard deviation. Equation 12.1 becomes:

$$F(u_1, u_2, \ldots, u_n) = 0. \qquad (12.6)$$

Equation 12.4 becomes:

$$p(safe\ region) = \frac{1}{(\sqrt{2\pi})^n} \iiint e^{-\frac{1}{2} \sum_{1}^{n} u_i^2} du_1\, du_2 \cdots du_n. \qquad (12.7)$$

The transformed standard space is shown in Figure 12.3. The trans-formed hypersurface F divides the multi-dimensional space into a safe and an unsafe region, with the values of F in the safe region equal to or smaller than 0, and those in the unsafe region greater than 0.

If *hyperspheres* are drawn in the transformed standard space, with cen-tres in the origin, the smallest hypersphere that touches the hypersurface covers a space that represents a close and conservative approximation of the entire safe region (see Figure 12.3). It should be noted that in the transformed standard space, the high probability values congregate in the region close to the origin (observe the integrand in Equation 12.7). This allows the integration of Equation 12.7 in the smallest hypersphere that touches the hypersurface to give a close and

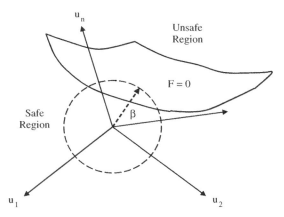

Figure 12.3 The transformed standard space where all transformed coordinates have a zero mean and a unit standard deviation. The cumulative probability in the smallest hypersphere that touches the hypersurface gives a conservative approximation of the total cumulative probability in the entire safe region. The radius of the hypersphere is the β reliability index.

conservative approximation of the total cumulative probability in the entire safe region.

The radius of any hypersphere, r, is equal to the square root of the sum of all parameter values that give that radius:

$$r = \sqrt{\sum_{1}^{n} u_i^2}. \qquad (12.8)$$

The smallest radius that touches the hypersurface is the β reliability index. If all the parameters associated with the hypersphere follow a standard normal distribution (with 0 means and unit standard deviations), the probability distribution of the hypersphere along its radius also follows that of a standard normal distribution, as can be seen by examining the integrand in Equation 12.7. The integration of Equation 12.7 in the smallest hypersphere that touches the hypersurface can be transformed into the following equivalent integration along the radius from 0 to β.

$$p(hypersphere) = \frac{2}{\sqrt{2\pi}} \int_0^\beta e^{-\frac{1}{2}r^2} dr \qquad (12.9a)$$

The above equation has the correct asymptotic value of 1 when β approaches ∞. Because of symmetry, the integration in Equation 12.9a from 0 to β can be changed to the following integration from $-\beta$ to β:

$$p(hypersphere) = \frac{1}{\sqrt{2\pi}} \int_{-\beta}^{\beta} e^{-\frac{1}{2}r^2} dr. \qquad (12.9b)$$

Furthermore, because of the congregation of the high probability values near the centre of the hypersphere, the integration in Equation 12.9b from $-\beta$ to β can be changed to the following integration from $-\infty$ to β. This makes the integration to be that of a cumulative standard normal distribution $\Phi(\beta)$.

$$p(hypersphere) = \frac{1}{\sqrt{2\pi}} \int_{-\infty}^{\beta} e^{-\frac{1}{2}r^2} dr = \Phi(\beta) \qquad (12.9c)$$

The probability of the safe region, therefore, is approximated by $\Phi(\beta)$; whereas the probability of the unsafe region is approximated by $1 - \Phi(\beta)$, which is also equal to $\Phi(-\beta)$ because of the symmetry of the distribution.

The search for the cumulative probability in the safe region becomes the search for the β reliability index. The β reliability index is the smallest radius in Equation 12.9c that touches the design hypersurface. Frantzich *et al.* (1997) used this method to define the fire safety design parameters that can meet a certain β reliability index value. They used an iterative method to define the design parameters, and hence the hypersurface, until the β reliability index reached a target value of 1.4, which gave a probability of success of 92 % or a probability of failure of 8 %. Hasofer and Qu (2002) showed a quicker way to find the β reliability index value if the design hypersurface can be simplified by regression into a quadratic equation. With the help of the Lagrange's method of undetermined multipliers, the root that gives the smallest radius gives the β reliability index.

We will show a simple example here to illustrate the β reliability index method. Suppose the design surface is governed by:

$$G = L - R + S = 0, \qquad (12.10)$$

where L is the fire load with a normal probability distribution $N(\mu_L, \sigma_L)$, R is the fire resistance with a normal probability distribution $N(\mu_R, \sigma_R)$, and S is the design safety margin with a fixed value. For example, the fire load, L, could be the equivalent standard fire time and the fire resistance, R, could be the FRR. Both of these parameters were discussed previously in Chapter 8. The design safety margin S is a parameter that can be used to allow a lower fire load, L, to cause a failure. It forces the fire resistance, R, to have a higher mean value in order to have the same probability of success in the safe region.

Equation 12.10 can be mapped into a standard space (Figure 12.3) using the Hasofer–Lind transformation:

$$u_L = \frac{L - \mu_L}{\sigma_L}; \quad u_R = \frac{R - \mu_R}{\sigma_R}. \tag{12.11}$$

Equation 12.10 becomes:

$$u_L - \frac{\sigma_R}{\sigma_L} u_R - \frac{\mu_R - \mu_L - S}{\sigma_L} = 0. \tag{12.12}$$

For this illustration, let us assume that this is a fire resistance problem with a design fire against a structural element which is constructed with a certain FRR. Both the design fire and the FRR are assumed to be stochastic parameters. This is different from the one that was discussed previously in Chapter 8, where only the design fire was a stochastic parameter. The design fire is assumed to have an equivalent standard fire time that is normally distributed $N(\mu_L, \sigma_L)$ and the structural element is assumed to have a FRR that is also normally distributed $N(\mu_R, \sigma_R)$. We further assume $\mu_L = 40$ minutes, $\sigma_L = 10$ minutes, $\mu_R = 60$ minutes, $\sigma_R = 10$ minutes and the design safety margin $S = 10$ minutes. With these assumed values, Equation 12.12 becomes:

$$u_L - u_R - 1 = 0. \tag{12.13}$$

To search for the β reliability index value, Equation 12.13 is plotted in a standard space in Figure 12.4. In a two-dimensional problem, the hypersurface is a curve, or a line as in this case. The β reliability index is the smallest radius that touches the line of Equation 12.13. The index value can be found by numerical iteration of the u_L and u_R values until the smallest radius is found.

For a two parameter problem, the β reliability index has an analytical solution:

$$\beta = \frac{\mu_R - \mu_L - S}{\sqrt{\sigma_R^2 + \sigma_L^2}}. \tag{12.14}$$

The analytical solution can be derived by differentiating Equation 12.8 and search for the minimum radius that satisfies Equation 12.10.

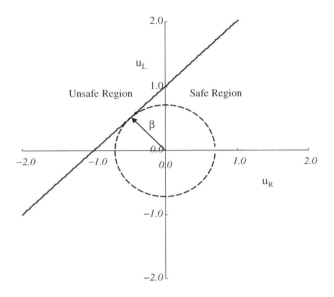

Figure 12.4 A plot of Equation 12.13 with the smallest radius that touches the line as the β reliability index.

For this simple problem, the β reliability index has a value of $1/\sqrt{2}$, or 0.707. The probability of success in the safe region is approximated by $\Phi(0.707)$ which has a value of 0.760. Conversely, the probability of failure in the unsafe region is approximated by $1 - \Phi(0.707)$, or $\Phi(-0.707)$, which has a value of 0.240.

Note that if the FRR in Equation 12.14 is a fixed value with no probability distribution, the problem becomes a one-dimensional problem. The design line becomes a point on the u_L axis with $\beta = (\mu_R - \mu_L - S)/\sigma_L$. The problem reverts back to the one that was discussed previously in Chapter 8.

12.3.2 Treatment of Parameters with Uncertain Values

For those parameters with uncertain values but without a known probability distribution, there are no standard treatment methods. The uncertain values need to be determined through research or survey. Failing that, the uncertain values need to be assumed and need to be agreed upon between fire safety engineers and authorities having jurisdiction. One way to check what value can be assumed is to do a *parametric study*. If the solution is not sensitive to the input value, any assumed value can be made. If the solution is very sensitive to the input value, then a more serious consideration is required to decide what value can be assumed.

One example of a parametric study is the one by Hadjisophocleous and Yung (1994) concerning the sensitivity of the reliability of fire protection systems on the expected risk to life in a building. Figure 12.5 is a reproduction of their figure on the effect of the reliability of a central smoke alarm system on the expected risk to life in a building which is either with or without sprinkler protection. The figure shows that the reliability of the central smoke alarm system has an effect on the relative expected risk to life when there is no sprinkler protection in the building, and a negligible effect when there is already sprinkler protection in the building. That means any assumed value of the reliability of the central smoke alarm is fine when there is sprinkler protection in the building, but a more careful consideration is required when there is no sprinkler protection in the building.

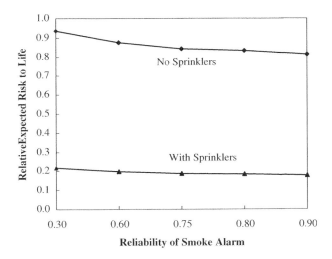

Figure 12.5 The reliability of a central smoke alarm system has an effect on the relative expected risk to life when there are no sprinklers in the building and a negligible effect when there are already sprinklers in the building   (from Hadjisophocleous and Yung, 1994, reproduced by permission of the International Association for Fire Safety Science).

12.4 Summary

Uncertainty exists in all disciplines, from quantum mechanics to financial markets. Uncertainties in fire safety design parameters give rise to uncertainties in fire safety designs and in turn uncertainties in fire risk assessment.

Fire safety parameters with uncertain values can be a result of many factors, but can be generally grouped into the following four major types. They are: (1) parameters with a probability distribution of different values, (2) parameters with random values, (3) parameters with unknown values and (4) parameters with future unknown values.

For parameters with a probability distribution, the probability of success or failure in fire safety designs can be analysed in a multi-dimensional space with the space divided by the fire safety design into a safe and an unsafe region. Two techniques were discussed that can be used to find the cumulative probability in the safe region. One is the Monte Carlo method which employs repeated random sampling to obtain the cumulative probability in the safe region. The other is the β reliability index method which uses the largest hypersphere that can fit into the safe region to find the cumulative probability in the safe region.

For those parameters with uncertain values but without a known probability distribution, there are no standard treatment methods. The uncertain values need to be determined through research or survey. Failing that, the uncertain values need to be assumed and agreed upon between fire safety engineers and regulators. One way to check what value can be assumed is to do a parametric study. If the solution is not sensitive to the input value, any assumed value can be made. If the solution is very sensitive to the input value, then a more serious consideration is required to decide what value can be assumed.

12.5 Review Questions

12.5.1 Redo the fire resistance example by increasing the fire resistance mean value, μ_R, from 60 to 70 minutes. Keep all other values the same. Calculate, by iteration as well as by using Equation 12.14, the β reliability index and the probability of success in the safe region. Compare the results with those in the example and see how the safe region has expanded.

12.5.2 Derive the analytical solution described by Equation 12.14. Take derivative of Equation 12.8 to find the minimum radius that satisfies Equation 12.10.

References

Frantzich, H., Magnusson, S.E., Holquist, B. and Ryden, J. (1997) Derivation of Partial Safety Factors for Fire Safety Evaluation Using the Reliability Index β Method. Proceedings of the Fifth International Symposium on Fire Safety Science, March 1997, Melbourne, pp. 667–78.

Hadjisophocleous, G.V. and Yung, D. (1994) Parametric Study of the NRCC Risk-Cost Assessment Model for Apartment and Office Buildings. Proceedings of the 4th International Symposium on Fire Safety Science, June 1994, Ottawa, pp. 829–40.

Hasofer, A.M. and Lind, N.C. (1974) Exact and invariant second-moment code format. *Journal of the Engineering Mechanics Division, ASCE*, **100** EM1, 111–21.

Hasofer, A.M. and Qu, J. (2002) Response surface modelling of monte carlo fire data. *Fire Safety Journal*, **37**, 772–84.

Notarianni, K.A. (2002) Uncertainty, *SFPE Handbook of Fire Protection Engineering*, 3rd edn, Section 5: Chapter 4, National Fire Protection Association, Quincy, MA.

13

Fire Risk Management

13.1 Overview

The assessment of expected occupant fatalities and property loss in a building for a particular fire scenario is achieved by modelling fire growth, smoke spread, fire spread, occupant evacuation and fire department response. Expected risk to life (ERL) for the occupants living in a building over the design life of the building is the sum of all expected occupant fatalities from all probable fire scenarios that may occur in a building over the design life of the building.

In the same way, the expected risk to property in the building over the design life of the building is determined by the sum of all expected property losses from all probable fire scenarios that may occur in a building over the design life of the building. The total expected fire cost (EFC) can be established if we add the initial capital cost of the fire protection measures and the maintenance cost of the fire protection measures over the design life of the building to the expected risk to property. The EFC represents the total cost of any fire safety design option to the building owner who has to pay for all costs including capital, maintenance and expected fire losses. As the owner is responsible for the EFC he or she will therefore be interested in establishing its lowest possible value.

The ability to assess the ERL and EFC values, as described in this book, allows the comparisons of the ERL and EFC values of different fire safety design options. Those fire safety design options that can provide equivalent or lower ERL values in comparison with that provided by the code-compliant fire safety design are considered acceptable alternative

Principles of Fire Risk Assessment in Buildings   D. Yung
© 2008 John Wiley & Sons, Ltd

design options. Those acceptable alternative design options that have the lowest EFC are the cost-effective alternative fire safety design options. Some examples of equivalent fire safety designs and cost-effective fire safety designs from the risk-cost assessment model fire risk evaluation and cost assessment model (FiRECAM) are discussed.

Regular inspection and maintenance of fire protection systems are required in risk-based, or performance-based, fire safety designs. Without such regular maintenance and evacuation drills, the consequence is that the ERL to the occupants is higher than that assumed by the fire safety design. The reliability of fire protection systems can be modelled based on failure rate and service time interval. The modelling equations are described and some examples are given.

13.2 Fire Risk Management

In Chapters 7–11, we discussed how expected occupant fatalities and property loss in a building for a particular fire scenario can be assessed by modelling of fire growth, smoke spread, fire spread, occupant evacuation and fire department response. The sum of all expected occupant fatalities from all probable fire scenarios that may occur in a building over the design life of the building gives the ERL for the occupants living in the building over the design life of the building. This summation is given in Equation 13.1, where P_i is the probability of occurrence of fire scenario i over the design life of the building, OF_i is the expected occupant fatalities for fire scenario i, and n is the total number of probable fire scenarios.

$$ERL = \sum_{i=1}^{n} P_i \, OF_i \qquad (13.1)$$

Similarly, the sum of all expected property losses from all probable fire scenarios that may occur in a building over the design life of the building gives the expected risk to property in the building over the design life of the building. Adding the initial capital cost of the fire protection measures and the maintenance cost of the fire protection measures over the design life of the building to the expected risk to property gives the total EFC.

The EFC represents the total cost of any fire safety design option to the building owner who has to pay for all costs including capital, maintenance and expected fire losses. Each fire safety design may have a different set of capital cost, maintenance cost and expected fire loss. For example, a fire safety design may have a higher initial capital cost, but lower maintenance cost and expected fire losses. Or a fire safety design may have a lower initial capital cost, but higher maintenance cost and expected losses. The EFC is the total cost that the owner is responsible and therefore interested in minimizing its value. This EFC is given in Equation 13.2, where P_i is the probability of occurrence of fire scenario i over the design life of the building, PL_i is the expected property loss for fire scenario i, and n is the total number of probable fire scenarios, CC is the initial capital cost of all fire protection measures, and MC is the present value of the maintenance cost of these fire protection measures over the design life of the building.

$$EFC = \sum_{i=1}^{n}(P_i\ PL_i) + CC + MC \qquad (13.2)$$

In this chapter, we will discuss fire risk management that involves the identification of fire safety design options that can provide a certain acceptable level of ERL. We will also discuss cost-effective fire risk management that involves not only the identification of fire safety design options that can provide the acceptable level of ERL, but also the lowest EFC. The ERL and the EFC depend on what fire protection measures are used in a fire safety design option, how well they work when fires occur, and what are the associated capital and maintenance costs. The ability to assess the ERL and the EFC allows us to compare the ERL and EFC values of different fire safety design options. Those fire safety design options that can provide equivalent or lower ERL values in comparison with that provided by the *code-compliant* fire safety design are considered acceptable alternative design options. Those acceptable alternative design options that have the lowest EFC are the *cost-effective* alternative fire safety design options.

We will also discuss in this chapter the need of ongoing inspection, maintenance and evacuation requirements for risk-based, or performance-based, fire safety designs. If a certain reliability is assumed for a fire protection system in the fire safety design, regular inspection and maintenance are required in order to maintain that level of reliability. If a certain evacuation performance is assumed in the fire safety design, regular evacuation training and drills are required in order to maintain that level of evacuation performance.

In this chapter, *reliability* is defined as the probability that the fire protection system will operate and perform its fire protection function as designed when a fire occurs. For example, sprinkler reliability is defined as the probability that the sprinkler will activate and will control the fire when a fire occurs.

13.3 Alternative Fire Safety Designs

In this section, we will look at some examples of *equivalent* fire safety *designs*. Equivalent fire safety designs are designs that can provide equivalent or lower ERL values in comparison with that provided by the code-compliant fire safety design. We will also look at some examples of cost-effective fire safety designs. Cost-effective fire safety designs are designs that can provide equivalent or lower ERL but with the lowest EFC.

As was discussed earlier in this chapter and throughout this book, the assessment of the ERL and EFC values requires the consideration of all probable fire scenarios that may occur in a building over the design life of the building and, for each fire scenario, the calculation of fire growth, smoke spread, fire spread, occupant evacuation and fire department response. The assessment of the ERL and EFC values, therefore, involves many calculations and the only practical way to do it is through the use of computer models. There are a few such comprehensive risk assessment models that have been developed in the world over the past 20 years. These computer models took many years and much resources to develop because of their complexities. Notable models include *FiRECAM*, the 'fire risk evaluation and cost assessment model' for apartment and office buildings that was developed at the National Research Council Canada (NRCC) in the 1990s (Yung, Hadjisopho-cleous and Proulx, 1997); *CESARE-RISK*, the risk and cost assessment model that was developed at the Victoria University of Technology in Australia in the 1990s (Zhao and Beck, 1997); and *FIERAsystem*, the fire risk assessment tool for light industrial buildings that was developed at the NRCC in the early 2000s (Benichou *et al.*, 2005).

The above models were developed with the modelling concepts that were discussed in this book. Of these three models, the only one that is available for use by the general public is the FiRECAM model which can be downloaded from the NRCC web site (FiRECAM, 2008). The author of this book was in charge of the development of the FiRECAM model when he was a fire researcher at the NRCC. The FiRECAM model was developed in collaboration with the development of the *CESARE-RISK* model at the Victoria University of Technology in Australia, headed by Professor Vaughan Beck, (Beck and Yung, 1990; Yung and Beck, 1995; Beck, 1997; Richardson, 2003). The FIERAsystem model was developed at the NRCC as an extension of the FiRECAM model for light industrial buildings, headed by Professor George Hadjisophocleous who was also a fire researcher at the NRCC but is now a professor at Carleton University in Canada.

In the following subsections, we will discuss examples of equivalent fire safety designs and cost-effective fire safety designs. We will use published case studies of FiRECAM as these examples.

13.3.1 Equivalent Fire Safety Designs

Equivalent fire safety designs are designs that provide equivalent, or lower, ERL to the occupants in a building in comparison with that from a code-compliant fire safety design. Equivalent fire safety designs can be building fire safety designs that provide equivalent life safety to the occupants in a building, or community fire safety regulations that provide equivalent life safety to the people in a community. In this example, we will discuss the case of community fire safety regulations that affect the life safety of the people in a community. In the next section when we discuss cost-effective fire safety designs, we will discuss the case of building fire safety designs that affect the life safety of the people in a building.

This example is taken from a previous FiRECAM case study by Benichou, Yung and Hadjisophocleous (1999). The case study discussed the issue whether to build a new fire station or mandate sprinkler protection for all buildings when a new community, or subdivision, is built. Such an issue involves both fire safety considerations as well as cost consideration. The case study addressed only the fire safety issue which will be discussed below.

In this case study, a three-storey apartment building was used to represent the building stock in a new community which could consist of single houses to high-rise buildings. The floor plan was assumed to be as shown in Figure 13.1. Eight apartment units were assumed per floor with 2.5 occupants being assumed per unit. Fire could start in any apartment unit. The ERL was calculated using FiRECAM for various *fire department travel times* and for two sprinkler options: with or without sprinkler protection. The various fire department travel times represent community management strategies whether to build a new fire station or to use existing fire stations for longer travels. In this study, sprinkler protection was assumed to have a 95 % reliability. The results are reproduced here in Figure 13.2.

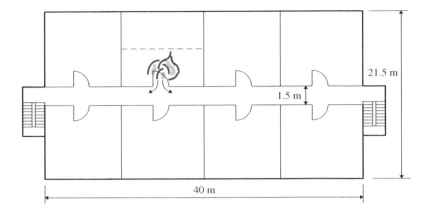

Figure 13.1 Floor plan of the three-storey model apartment building that was used for the study (from Benichou, Yung and Hadjisophocleous, 1999, reproduced by permission of Interscience Communications Ltd).

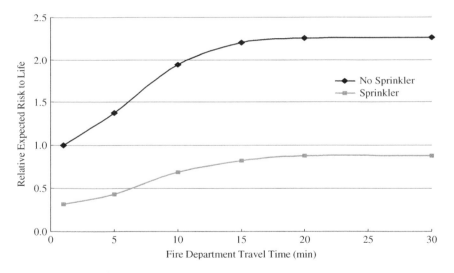

Figure 13.2 Relative expected risk to life as a function of fire department travel time and with and without sprinkler protection (from Benichou, Yung and Hadjisophocleous, 1999, reproduced by permission of Interscience Communications Ltd).

In Figure 13.2, the ERL of the mandatory sprinkler option is plotted, for comparison purposes, as a relative value of that of the no sprinkler option. The results show that the ERL of the mandatory sprinkler option is always better than that of the no sprinkler option, no matter how fast the fire department travel time is. This result is not unexpected because sprinkler option provides faster on-site fire suppression than any fire department response can provide. This result suggests mandatory sprinkler protection provides better fire safety to the people in a new community than building a new fire station in a new community. Cost consideration is a different matter which is not discussed here. For more details on this case study, consult the paper (Benichou, Yung and Hadjisophocleous, 1999).

The case study is used here to demonstrate the use of fire risk assessment to show whether alternative fire safety strategies can provide equivalent or better fire safety. Without fire risk assessment, it would be difficult to argue whether one fire safety strategy is better than another.

13.3.2 Cost-effective Fire Safety Designs

Cost-effective fire safety designs are designs that provide equivalent, or lower, ERL to the occupants in a building than that provided by a code-compliant fire safety design, but with the lowest EFC. In this example, we will discuss cost-effective fire safety design options for a large office building.

The example is taken from a previous published FiRECAM case study by Yung, Hadjisophocleous and Yager (1998). The case study was one of four case studies that were invited by the *2nd International Conference on Performance-Based Codes and Fire Safety Design Methods* in 1998 to present alternative fire safety design options that can meet both the building code's fire safety requirements and the building owner's expectation of low cost. The model building is a conference-specified 40-storey office building assumed to be located in the country that participated in the case studies.

For the FiRECAM case study, the building was assumed to be located in Canada. Any proposed fire safety design options were required to meet both the National Building Code of Canada's (NBCC) fire safety requirements and the building owner's expectation of lowest fire protection cost and losses possible. Since NBCC did not have performance requirements, the case study was to consider alternative fire safety design options that could provide the occupants with the same, or better, level of fire safety as implied by the prescriptive requirements of the NBCC. In addition to meeting the building code requirements and to have the lowest fire protection cost and losses, the building owner would like to have a refuge area on each floor for occupants with disabilities.

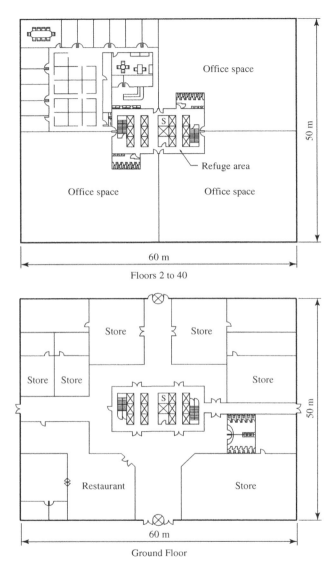

Figure 13.3 Floor plans selected for the case study (from Yung, Hadjisophocleous and Yager, 1998, reproduced by permission of the Society of Fire Protection Engineers).

The floor plan that was chosen by Yung, Hadjisophocleous and Yager (1998) for the case study is reproduced here in Figure 13.3. The offices were arranged around the building perimeter to allow maximum use of the window areas. Service elevators, stairs and washrooms were

Table 13.1 Five fire safety design options considered for the case study.

Design option	Fire resistance rating (min)	Refuge area	Sprinklers (reliability %)
Reference	120	No	95
2	90	No	95
3	90	Yes	95
4	90	No	99
5	90	No	No

placed in the centre core. A refuge area, protected from fire and smoke, was provided in the centre core for occupants with disability to stay and wait for rescue.

Five fire safety design options were considered by Yung, Hadjisophocleous and Yager (1998) for this case study. They are listed in Table 13.1. The five design options include the code-compliant design as the reference design. The other four design options include lowering the fire resistance rating, the provision of refuge area and various levels of sprinkler protection. All five options have a central alarm system with voice communication which is required by the NBCC.

The calculated ERL of the five design options are reproduced here in Figure 13.4 from Yung, Hadjisophocleous and Yager (1998). The ERL values are plotted relative to the reference design option which is the

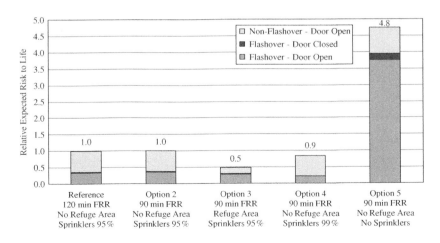

Figure 13.4 Relative expected risk to life for the five design options shown in Table 13.1 (from Yung, Hadjisophocleous and Yager, 1998, reproduced by permission of the Society of Fire Protection Engineers).

code-compliant design option. Also shown for each ERL value are the contributions by the fire type in the compartment of fire origin and the door condition (review Chapter 7). It is not surprising that the major contributors to each ERL are the severe flashover fire with the door open followed by the nonflashover fire with the door open.

Figure 13.4 shows that lowering the fire resistance rating from 120 to 90 minutes (Option 2) does not have any measurable impact on the ERL; but removing the sprinkler protection (Option 5) can increase the ERL value by a factor of 4.8. The other two options, Options 3 and 4, provide lower ERL values. Option 3 has a refuge area on each floor that can provide additional protection, especially for people with disability; whereas Option 4 has a more reliable sprinkler system that can provide better fire suppression. Option 2, 3 and 4 are therefore equivalent fire safety design options to the code-compliant reference design. The question is which of these equivalent design options has the lowest EFC. The one that has the lowest EFC is the cost-effective fire safety design.

The calculated EFC for the five fire safety design options are reproduced here in Figure 13.5 from Yung, Hadjisophocleous and Yager (1998). Also shown for each EFC are the breakdown of the capital cost

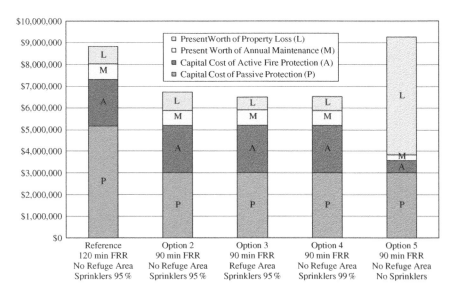

Figure 13.5 Expected fire cost for the five design options shown in Table 13.1 (from Yung, Hadjisophocleous and Yager, 1998, reproduced by permission of the Society of Fire Protection Engineers).

of passive protection (P), the capital cost of active protection (A), the present worth of annual maintenance cost (M) and the present worth of expected fire loss (L). Figure 13.5 shows that Option 5, with a lower fire resistance rating and without sprinkler protection, has a slightly higher EFC than that of the code-compliant reference design. Even though this option has a lower capital cost for both passive and active fire protections, it has a much larger expected fire loss that costs more than the savings in capital cost. The other three options all have lower EFC than that of the reference design. The saving comes mainly from lower capital cost for lower fire resistance rating. Options 3 and 4 have slightly lower EFC values than that of Option 2 because they have lower expected fire losses. The lower fire losses are the result of their having additional fire protections which are the refuge areas and the more reliable sprinkler system. Options 3 and 4 provide not only the lowest ERL but also lowest EFC. Options 3 and 4 are, therefore, the cost-effective fire safety design options.

For more details on this case study, refer to the paper by Yung, Hadjisophocleous and Yager (1998). The main objective of the discussion of this example is to show that comprehensive risk assessment models can provide not only risk assessment but also cost assessment. This allows comparisons of alternative fire safety design options to see whether they can provide the required level of fire safety but also whether they have the lowest fire protection cost and expected fire loss. The ultimate goal for all concerned is to find not only equivalent fire safety designs, but more importantly, cost-effective fire safety designs.

13.4 Impact of Inspection and Maintenance on System Reliability

The *inspection* and *maintenance* of installed fire protection systems are an integral part of fire risk management. Without regular inspection and maintenance, the installed fire protection systems may not work as reliably as intended nor as well as designed. This is true for all building systems, such as heating and air conditioning. Regular inspection and maintenance is the key to good reliability and performance.

Regular inspection and maintenance of fire protection systems can be a mandatory requirement in performance-based fire safety designs. If certain reliabilities are assumed for the fire protection systems, they must be backed up by regular, documented, inspection and maintenance. Without such highly regimented inspection and maintenance, the

214 Fire Risk Management

assumed reliabilities are not assured. The consequence is that the ERL to the occupants is higher than that assumed by the fire safety design.

It should be noted that fire protection systems are required to go through commissioning tests before they can go into service. The commissioning tests ensure that the fire protection systems are installed properly and work properly (Elovitz, 2006). Once they go into service, these systems may still fail over time because system components have limited service lives. Regular inspection and maintenance, therefore, are needed to locate and remove malfunctioned components before their service is required in a fire situation.

Fire protection systems, therefore, depend on good engineering design and analysis to come up with the right systems that work effectively, commissioning tests to confirm that they work as they are supposed to, and an adequate maintenance schedule to ensure that they work reliably.

13.4.1 Component Reliability

A component's reliability depends on the product of its failure rate λ (frequency/time) and its time in service t. The product of λ and t is a nondimensional parameter. Various functional relationships are employed to model the dependence of reliability on this nondimensional parameter λt (Modarres and Joglar-Billoch, 2002). One simple relationship that is often used is the following exponential relationship.

$$P_R[C_i] = e^{-\lambda_i t} \qquad (13.3)$$

In the above equation, $P_R[C_i]$ is the reliability of component C_i, λ_i is the failure rate (frequency/time) of component C_i, and t is the time in service. Equation 13.3 is plotted in Figure 13.6. The reliability is shown to decrease with the increase in λt. It also has the correct limiting value of 1 when λt is 0 and a value of 0 when λt is large.

Let us look at an example of the reliability of sprinkler protection. The failure rates of sprinkler components can range from 2×10^{-7} to $5 \times 10^{-4}\,h^{-1}$, with many having a value close to $5 \times 10^{-6}\,h^{-1}$ (Fong, 2000). If these components are inspected and maintained annually, the service time is $8.76 \times 10^3\,h$. This gives λt a value of 4.38×10^{-2} and, from Equation 13.3, a reliability value of 95.7%. If these components are not inspected and maintained, similar calculations show that the reliability drops to 80.3% in five years' time and 64.5% in ten years' time. This shows the importance of regular inspection and maintenance in order to achieve high reliability. Without regular inspection and

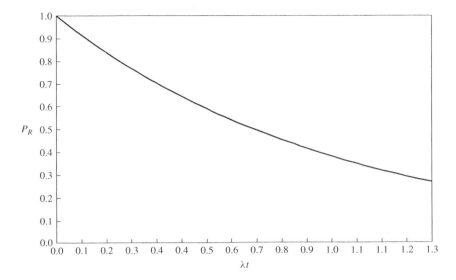

Figure 13.6 Component reliability P_R as a function of the nondimensional parameter λt, where λ is the failure rate (frequency/time) and t is the time in service.

maintenance, the reliability can not be assured. The best maintenance is a built-in supervised system which monitors all components continuously and can detect any malfunction immediately (Dungan, 2007).

13.4.2 System Reliability

A system's reliability depends on the reliabilities of its components. The reliability is usually analysed using a method called the *fault tree analysis*. In a fault tree analysis, the system components are grouped together based on how they work together. The assembled components look like a tree with the basic components in the bottom, the subsystems in the mid section, and the final system at the top. The fault tree analysis is a bottom-up analysis. The failure of any component at the bottom will lead to the failure of a subsystem in the mid section; and the failure of any subsystem will lead to the failure of the whole system at the top.

The failure of a subsystem depends on how the components work together. If each component is critical to the success of the subsystem, then any failure will lead to the failure of the subsystem. On the other hand, if a backup component is used, then both components have to fail before the subsystem fails. These two working relationships also apply to how subsystems work together. To distinguish the flow of failure

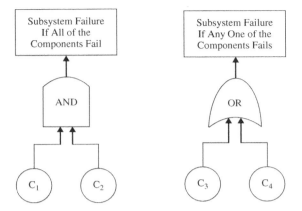

Figure 13.7   Logic gates 'AND' and 'OR' that are used in fault tree analysis.

information up the tree from these two different types of groupings, *logic gates* of '*AND*' and '*OR*' are used which are shown in Figure 13.7. The 'AND' gate allows failure information to go up the tree if all the components fail; whereas the 'OR' gate allows failure information to go up the tree if any one of the components fails.

In an 'AND' gate situation, the failure of the subsystem depends on the failure of all its components. The probability of failure of each component, $P_F[C_i]$, is the complement of the reliability of each component, $P_R[C_i]$, as expressed in the following:

$$P_F[C_i] = 1 - P_R[C_i] = 1 - e^{-\lambda_i t}. \tag{13.4}$$

The probability of failure of the subsystem, $P_F[AND]$, is the product of the probabilities of failure of all its components, as expressed in the following:

$$P_F[AND] = \prod_{i=1}^{n} P_F[C_i] = \prod_{i=1}^{n} (1 - e^{-\lambda_i t}). \tag{13.5}$$

The reliability of the subsystem, $P_R[AND]$, is the complement of its probability of failure, $P_F[AND]$, as expressed in the following:

$$P_R[AND] = 1 - P_F[AND] = 1 - \prod_{i=1}^{n} (1 - e^{-\lambda_i t}). \tag{13.6}$$

Note that in an 'AND' gate situation, the reliability of the subsystem, as expressed by Equation 13.6, is higher than that of each component.

Using the same failure rate that was used earlier for a single component, $5 \times 10^{-6}\,h^{-1}$, the same annual inspection and maintenance with a service time of $8.76 \times 10^3\,h$, and assuming all components have the same failure rate, the reliability of a two component system, obtained from Equation 13.6, is 99.8 %. This is higher than the reliability obtained earlier for a single component, 95.7 %. The higher reliability is expected because the main objective of a backup system is to increase system reliability.

In an 'OR' gate situation, which is the opposite of the 'AND' gate situation, the reliability of the subsystem depends on the reliabilities of all its components. The reliability of the subsystem, $P_R[OR]$, is the product of the reliabilities of all its components, as expressed in the following:

$$P_R[OR] = \prod_{i=1}^{n} P_R[C_i] = e^{-(\sum_{i}^{n} \lambda_i)\, t}. \qquad (13.7)$$

Note that in an 'OR' gate situation, the reliability of the subsystem, as expressed by Equation 13.7, is lower than that of each component. Using the same failure rate that was used earlier for a single component, $5 \times 10^{-6}\,h^{-1}$, the same annual service time of $8.76 \times 10^3\,h$, and assuming all components have the same failure rate, the reliability of a two component system, obtained from Equation 13.7, is 91.6 %. This is lower than the reliability obtained earlier for a single component, 95.7 %. The lower reliability is expected because the probability of failure is higher when there are more components that can fail than a single component.

13.4.3 Impact of System Reliability on Expected Risk to Life

The impact of fire protection systems on the ERL and the EFC were discussed earlier in this chapter. How much the impact is also depends on the reliability of the fire protection system. Reliability affects the fire scenarios that are considered in the assessment of the ERL and EFC values. As was discussed earlier in this chapter and throughout this book, the assessment of the ERL and EFC values requires the consideration of all probable fire scenarios that may occur in a building over the design life of the building and, for each fire scenario, the calculation of fire growth, smoke spread, fire spread, occupant evacuation and fire department response.

We will look at an example of the impact of the reliability on the ERL. We will look at a published case study of FiRECAM as this

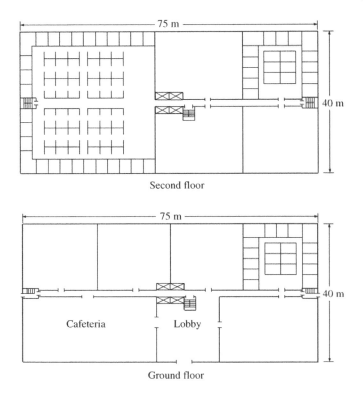

Second floor

Ground floor

Figure 13.8 Floor plans of four-storey office building (from Yung and Had-jisophocleous, 1997, reproduced by permission of the Fire Protection Research Foundation).

example (Yung and Hadjisophocleous, 1997). The case study used a typical four-storey office building with typical fire protection systems to study the effect of the reliability of sprinkler and central alarm systems on the ERL. The floor plans used in their study are reproduced here in Figure 13.8.

The results of the impact of *reliability* of *sprinkler* and *central alarm* systems on the ERL are reproduced in Figure 13.9. It should be noted that the results are only applicable to the building, the occupants and the fire protection systems that were assumed in this case study. The results, however, can still be used as an example to show the impact of the reliability of fire protection systems on the ERL. Figure 13.9 shows the relative ERL for various reliability values of central fire alarms and automatic sprinklers. The reference case is the one with no sprinkler protection and an alarm reliability of 80 %. Figure 13.9 shows that,

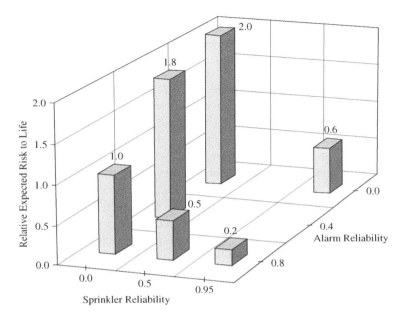

Figure 13.9   Relative expected risk to life for various reliability values of central fire alarms and automatic sprinklers (from Yung and Hadjisophocleous, 1997, reproduced by permission of the Fire Protection Research Foundation).

without sprinkler protection, the relative ERL doubles from 1.0 to 2.0 as the alarm reliability drops to zero. With sprinkler protection and at a reliability of 95 %, the relative ERL drops to 0.2 if the alarm reliability is at 80 %. With sprinkler protection and at a reliability of 95 %, the relative ERL still drops to 0.6 even if the alarm reliability drops to zero.

One interesting point to note is the comparison of the impact of central alarms and the sprinkler protections. Without the sprinkler protection, a central alarm with a reliability of 80 % provides a relative ERL of 1 (the reference case). On the other hand, without the central alarm (i.e. reliability at 0 %), a sprinkler protection with a reliability of 95 % provides a lower relative ERL of 0.6. This shows, as expected, a sprinkler system with a typical reliability provides a better safety than a central alarm with a typical reliability. This is especially true if the reliability of alarms can not be maintained properly due to nuisance false alarms and the possibility of disconnection by the occupants.

13.5 Impact of Evacuation Drills on Early Occupant Response and Evacuation

As was discussed in Chapter 10, regular evacuation training and drills are required in order to minimize the *delay start time* and to shorten the *movement time*. Regular evacuation training and drills allow the occupants to plan ahead, to quickly recognize the warning signals, and to expedite the pre-movement activities. Regular evacuation training and drills also allow the occupants to shorten the *movement time* to a safe place because it provides the occupants with prior knowledge of the best evacuation route to take. Regular evacuation training and drills are also important for the security staff. This allows them to quickly issue proper warnings and instructions to the occupants.

Regular evacuation training and drills are especially important for high-rise buildings where controlled selective evacuation of only certain floors is used rather than the uncontrolled total evacuation of the whole building. Example of controlled selective evacuation is to evacuate all the floors above the fire floor and only one floor below the fire floor. This helps to avoid congestion in the stairs and the slow down of the evacuation process. However, controlled selective evacuation only works if the evacuation instructions are clear and the occupants are willing to follow the instructions.

Regular evacuation training and drills can be a mandatory requirement in performance-based fire safety designs. If certain quick occupant response time and movement time are assumed, they must be backed up by regular, documented, successful evacuation training and drills. Without such highly regimented evacuation training and drills, the assumed quick occupant response time and movement time are not assured. The consequence is that the ERL to the occupants is higher than that professed by the fire safety design.

13.6 Summary

Expected occupant fatalities and property loss in a building for a particular fire scenario are assessed by modelling of fire growth, smoke spread, fire spread, occupant evacuation and fire department response. The sum of all expected occupant fatalities from all probable fire scenarios that may occur in a building over the design life of the building gives the ERL for the occupants living in the building over the design life of the building.

Similarly, the sum of all expected property losses from all probable fire scenarios that may occur in a building over the design life of the building gives the expected risk to property in the building over the design life of the building. Adding the initial capital cost of the fire protection measures and the maintenance cost of the fire protection measures over the design life of the building to the expected risk to property gives the total EFC. The EFC represents the total cost of any fire safety design option to the building owner who has to pay for all costs including capital, maintenance and expected fire losses. The EFC is the total cost for which the owner is responsible and therefore he or she will be interested in its lowest possible value.

Fire risk management involves the identification of fire safety design options that can provide a certain acceptable level of ERL. Cost-effective fire risk management involves not only the identification of fire safety design options that can provide the acceptable level of ERL but also the lowest EFC. The ability to assess the ERL and EFC values, as described in this book, allows the comparisons of the ERL and EFC values of different fire safety design options. Those fire safety design options that can provide equivalent or lower ERL values in comparison with that provided by the code-compliant fire safety design are considered acceptable alternative design options. Those acceptable alternative design options that have the lowest EFC are the cost-effective alternative fire safety design options. The assessment of the ERL and EFC values involves many calculations and the only practical way to do it is through the use of computer models. Some examples of equivalent fire safety designs and cost-effective fire safety designs from the risk-cost assessment model FiRECAM were discussed.

Regular inspection and maintenance of fire protection systems are required in risk-based, or performance-based, fire safety designs. If certain reliability is assumed for a fire protection system, regular inspection and maintenance are required in order to maintain that level of reliability. Similarly, regular evacuation training and drills are required in order to maintain that level of evacuation performance that has been assumed in the fire safety design. Without such regular maintenance and evacuation drills, the consequence is that the ERL to the occupants is higher than that assumed by the fire safety design. The reliability of fire protection systems can be modelled based on failure rate and service time interval. The modelling equations were described and some examples were given.

13.7 Review Questions

13.7.1 Explain why in Figure 13.4 flashover and nonflashover fires with the door of the compartment of fire origin open are major contributors to the ERL. (Review Chapter 7 and Chapter 9.)

13.7.2 Explain why sprinklers have significant impact on lowering the risk from flashover fires. (Review Chapter 7.)

13.7.3 Calculate the reliability of a fire protection system with five components if they are inspected and maintained annually. The fire protection system depends on all five components working (i.e. 'OR' gate situation). Assume each component has the same failure rate of $5 \times 10^{-6}\,h^{-1}$.

13.7.4 Repeat the above calculation if the system is inspected and maintained every three months.

References

Beck, V. (1997) Performance-based Fire Engineering Design and its Application in Australia, Proceedings of the Fifth International Symposium on Fire Safety Science, Melbourne, Australia, March 1997, pp. 23–40.

Beck, V.R. and Yung, D. (1990) A cost-effective risk assessment model for evaluating fire safety and protection in Canadian apartment buildings. *Journal of Fire Protection Engineering*, 2(3), 65–74.

Benichou, N., Kashef, A.H., Reid, I. *et al.* (2005) FIERAsystem: a fire risk assessment tool to evaluate fire safety in industrial buildings and large spaces. *Journal of Fire Protection Engineering*, 15(3), 145–72.

Benichou, N., Yung, D. and Hadjisophocleous, G.V. (1999) Impact of Fire Department Response and Mandatory Sprinkler Protection on Life Risks in Residential Communities, Proceedings of Interflam '99, Edinburgh, Scotland, July 1999, pp. 521–32.

Dungan, K.W. (2007) Reliability of fire alarm systems, *Fire Protection Engineering Magazine*, Society of Fire Protection Engineers, Bethesda, MD, Winter 2007 edition, pp. 38–48.

Elovitz, K.M. (2006) Commissioning smoke control systems, *Fire Protection Engineering Magazine*, Society of Fire Protection Engineers, Bethesda, MD, Fall 2006 edition, pp. 28–40.

FiRECAM (2008) *Fire Risk Evaluation and Cost Assessment Model*, National Research Council Canada, http://irc.nrc-cnrc.gc.ca/fr/frhb/firecamnew_e.html

Fong, N.K. (2000) Reliability study on sprinkler system to be installed in old high-rise buildings. *International Journal on Engineering Performance-Based Codes*, 2(2), 61–67.

Modarres, M. and Joglar-Billoch, F. (2002) Reliability, *SFPE Handbook of Fire Protection Engineering 2002*, 3rd edn, Section 5: Chapter 3, National Fire Protection Association, Quincy, MA.

Richardson, J.K. (ed) (2003) *History of Fire Protection Engineering*, National Fire Protection Association, Quincy, MA, pp. 274–75.

Yung, D. and Beck, V.R. (1995), Building fire safety risk analysis, *SFPE Handbook of Fire Protection Engineering 1995*, 2nd edn, Section 5: Chapter 11, National Fire Protection Association, Quincy, MA.

Yung, D. and Hadjisophocleous, G.V. (1997) Assessment of the Impact of Reliability of Fire Alarms and Automatic Sprinklers on Life Safety in Buildings, Proceedings of the 2nd Fire Risk and Hazard Assessment Research Application Symposium, San Francisco, California, June 25–27, 1997, pp. 132–41.

Yung, D., Hadjisophocleous, G.V. and Proulx, G. (1997), Modelling Concepts for the Risk-Cost Assessment Model FiRECAM and Its Application to a Canadian Government Office Building, Proceedings of the Fifth International Symposium on Fire Safety Science, Melbourne, Australia, March 1997, pp. 619–30.

Yung, D., Hadjisophocleous, G.V. and Yager, B. (1998) Case Study: The Use of FiRECAM™ to Identify Cost-Effective Fire Safety Design Options for a Large 40-Storey Office Building, Proceedings of the 1998 Pacific Rim Conference and 2nd International Conference on Performance-Based Codes and Fire Safety Design Methods, May 3-9, 1998, Maui, Hawaii, pp. 441–52.

Zhao, L. and Beck, V. (1997) The Definition of Scenarios for the CESARE-RISK Model, Proceedings of the Fifth International Symposium on Fire Safety Science, March 1997, Melbourne, Australia, pp. 655–66.

Index

Printed and bound by CPI Group (UK) Ltd, Croydon, CR0 4YY

26/01/2023

03184479-0002